D0384731

The Physics of Everyday Things

The Physics of
Everyday Things

The Extraordinary Science
Behind an Ordinary Day

James Kakalios

CROWN
NEW YORK

Library of Congress Cataloging-in-Publication Data
Names: Kakalios, James, 1958– author.
Title: The physics of everyday things : the extraordinary science
    behind an ordinary day / James Kakalios.
Description: First edition. | New York : Crown Publishers, [2017] |
    Includes bibliographical references.
Identifiers: LCCN 2016046355 | ISBN 9780770437732 (hardcover)
    | ISBN 0770437737 (hardcover) | ISBN 9780770437749 (ebook) |
    ISBN 0770437745 (ebook)
Subjects: LCSH: Physics—Miscellanea. | Science in popular
    culture.
Classification: LCC QC75 .K175 2017 | DDC 530—dc23 LC record
    available at https://lccn.loc.gov/2016046355

ISBN 978-0-7704-3773-2
Ebook ISBN 978-0-7704-3774-9

PRINTED IN THE UNITED STATES OF AMERICA

Illustrations by Peter Arkle, except on page 246 by Gene Ha
Jacket design by Tal Goretsky
Front jacket photographs (toaster, bread): GS/Gallery Stock

10 9 8 7 6 5 4 3 2 1

First Edition

To

Geoff and Camille Nash

and

Augusta Peterson

for

showing me that every day

can be extraordinary

CONTENTS

The Physics of Everyday Things

CHAPTER ONE

You Begin Your Day

It is early morning, and you're asleep in bed. Your slow, regular breathing and steady pulse mark the passage of time, bringing you closer to when you must get up and begin your day. Today will be a busy one, with a visit to the doctor followed by a flight to another city for a business presentation. The vintage clock on your wall, a gift from your grandmother, provides a comforting tick, tock *as the small bob hanging from the body swings rhythmically back and forth. Although the clock keeps good time, you rely on the alarm setting of your smartphone to wake you. But the first sensation that will register the start of your day will not be your hearing; it will be your sense of smell. Last night you set the* **digital**

timer *on your coffeemaker to start its brewing cycle ten min-*
utes before your phone's alarm will go off. Your room soon
fills with the aroma of fresh coffee, and you begin to stir.

The elegant physics of an oscillating pendulum
underlies the working of both the clock on the
wall and the electronic timer on your coffeemaker, and
plays a crucial role in many of the devices you will use as
you prepare for the day.

A pendulum is a very simple device, consisting of a
string, fixed at one end, with a mass, termed the "bob,"
attached at the other end. The oscillations of the pendulum bob provide visual confirmation of one of the most
important concepts in physics, that of the principle of
conservation of energy: "kinetic energy," the energy of
motion, can only be converted to "potential energy" (the
energy associated with a force acting on an object and
the distance over which that force can cause motion) and
vice versa. In a pendulum, you can increase the potential
energy of the mass on the string by lifting it up, rotating the bob to a higher level while keeping the string
taut, doing work against the gravity that pulls down on
the bob. Once you release the bob, its potential energy
is converted into kinetic energy as it moves in an arc of
a semicircle. As the bob swings to the other side, the
kinetic energy is converted back into potential energy.
Both the starting height and the final height at the other

end of the arc are the same—when you release the mass and don't push it, it can never rise to a greater height than where it started.

A pendulum is useful for keeping time. The time it takes for the bob to complete a full cycle as it swings back and forth does not depend on how heavy the weight is, or on how high the mass is lifted to start it swinging (at least, for relatively small excursions back and forth). The greater the height of the mass, the larger the arc as it swings back and forth, and the larger the kinetic energy and speed it will have at the bottom of its arc. The longer distance and the faster speed exactly balance out, so that the time it takes to complete a cycle is the same— regardless of how high the bob is raised. The only factor that controls the time for a cycle is the length of the string. A pendulum whose string is just a little less than ten inches long will take one second to complete a full oscillation. As it swings, some of the kinetic energy of the bob is transferred to the surrounding air, pushing the molecules out of the bob's way. A careful audit will find that the gain in kinetic energy of the air is exactly equal to the reduction of the total energy of the pendulum, which is why mechanical clocks—grandfatherly and otherwise—need periodic winding.

It's as true for the digital timer on the coffeemaker as it is for the mechanical pendulum—to mark the passing of time, one needs a power supply (as everything, even counting seconds, requires a source of energy) and a way to convert that energy into a periodically varying cycle. The coffeemaker is plugged into an outlet connected to an external electric power grid. Conveniently for us, the mechanism by which electric power is generated at a power plant automatically leads to an electric current that oscillates back and forth like a pendulum that can be exploited when making a timer.

Your electric company rotates coils of wire between the poles of large electromagnets, and to see how that leads to an alternating electric current, let's return to the simple mechanical oscillating pendulum. Let the bob at the end of the string have an electric charge, say from a few extra electrons sitting on it. Even if this pendulum has a frictionless pivot point and is swinging in a perfect vacuum, with no air drag, it will eventually slow down and come to rest. Where did the bob's energy go? Into electromagnetic waves, demonstrating a profound symmetry between electric and magnetic fields that will be exploited repeatedly throughout your day.

An "electric current" is defined as the motion of electric charges moving together, and as the electrically charged bob swings back and forth, changing its speed, it is a constantly changing current. The current is large at the bottom of the arc, when the bob is moving at its fastest, and the current is zero at the top of the arc, when the bob is momentarily stationary. Moving electric charges, as in a current, generate a magnetic field (this is known as Ampere's law); the faster they move,

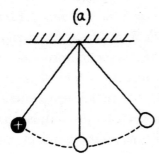

(a)

No Kinetic Energy
Large Potential Energy
No Current

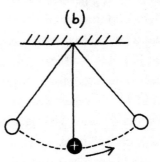

(b)

Large Kinetic Energy
No Potential Energy
Large Current

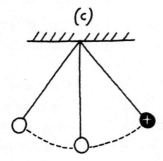

(c)

No Kinetic Energy
Large Potential Energy
No Current

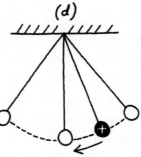

(d)

Some Kinetic Energy
Some Potential Energy
Changing Current

FIGURE I

the larger the magnetic field. The swinging bob, creating a constantly changing current, generates an equally varying magnetic field. In turn, this changing magnetic field generates a varying electric field (known as Faraday's law). This rhythmic oscillation of electric and magnetic fields is termed an "electromagnetic wave," which will have the same frequency as that of the oscillating bob. These waves carry energy, and thus it takes energy to create them. This is why the oscillations of an electrically charged bob will slowly die out, as its energy of motion is converted into electromagnetic waves. We could see these electromagnetic waves with the naked eye if the pendulum were swinging back and forth very rapidly (say, a thousand trillion times a second), in which case these waves would appear as visible light.

The power company employs the basic physics of electromagnetism when it generates the electric voltage available from the wall outlet, using coils of wire rotating between the poles of a magnet. The voltage provided by your electric company alternates smoothly from a positive voltage to a negative voltage, and back again, forming a wave that is mathematically identical to the variation in position of the pendulum bob as it oscillates back and forth, and is a natural consequence of how the electricity is produced. (This is why our electric power is called AC, for "alternating current.") The power plant is applying Faraday's law, which describes how a changing magnetic field will generate a voltage. As the coil turns, the magnitude of the magnetic field passing through the circular area of the coil varies, and a voltage is generated that sets up a current in the coil.* Think of the coil as a spool of thread with a very large diameter. When the area of the coil is facing the poles of the magnet, most of the magnetic field passes through it (along the length of the spool), but when it rotates by ninety degrees, hardly any of the field passes through the coil's area. A uniform rotation speed yields a smoothly varying voltage that changes back and forth in time, just like the motion of the pendulum bob. In the United States, the coils rotate sixty times a second, which is the frequency of the alternating voltage that is generated.

The fact that the voltage in the wall outlet varies smoothly back and forth sixty times a second means that it takes only 0.0167 second to complete one cycle. To slow this period down to one second, the coffeemaker's timer uses specially designed chips that shift the frequency of the alternating voltage.** One chip divides the incoming frequency by ten, so a voltage wave that oscillates sixty times a second now does so six times a second. Another

---

* High-pressure steam is used to turn the coil, and the steam is generated by boiling water via the burning of coal or natural gas or biomatter, or through the energy of nuclear reactions. Regardless of the fuel they use, all power plants use the same physics to generate electricity.

** These chips generate beats by adding a second frequency to the first signal (a process called "heterodyning"). The result is two oscillations, one at a higher frequency that is the sum of the two, and another at a lower frequency that is their difference. Using a filter, one can select for the lower frequency.

chip divides this frequency by an additional factor of six, so the frequency of six cycles per second is reduced to one cycle per second. This slower voltage wave is sent to another chip, which counts the number of times the voltage has its largest positive value (equivalent to watching how often the pendulum bob returns to its original starting position). This "counting" chip monitors the passing seconds, and with a little amount of additional circuitry, this information can be displayed on a digital clock. When you set the timer on your coffeemaker, you are instructing a chip to monitor this counting chip, and when the sum reaches a certain value (the time you specified for the coffeemaker to turn on), it sends another voltage to another part of the electronic system. This voltage is the same as the one created when you press the ON switch manually, and the brewing process begins.

The system for measuring time begins when we plug in the coffeemaker and set the correct time. If the coffeemaker is unplugged, then this preset is lost. So how does an electronic timer work when it is not connected to the external alternating-current power from the wall outlet?

*The coffee vapors waft into your room and are recognized by your still-not-fully-awake mind. In addition to setting the timer on the coffeemaker last night, you set the **alarm clock** on your smartphone. The alarm goes off, playing a preset tune stored in the phone's memory chip. You grumble as you* *check the time, because it is earlier than you normally need to get up. You are tempted to tap the alert marked SNOOZE. But as you inhale the coffee aroma, you notice your packed overnight bag sitting in the corner of your bedroom. Reminding yourself that you have a long day in front of you, you force yourself out of bed. Getting to your feet, you wince slightly as you put weight on your left foot. It will be good to get that looked at today.*

The problem of keeping time in a device not connected to an external electrical power supply is an old one; in fact, it predates the existence of electrical power. Old-fashioned alarm clocks used springs, and when the hands on the face of the clock reached a set point, a lever would be flipped, releasing another coiled spring. This other spring would then oscillate a striker bar between two metal shells, creating a clanging noise loud enough to wake the dead. The alarm in your smartphone is smaller, and the wake-up tune is less jarring, but the principle behind its operation is essentially the same.

Your smartphone uses something called a piezoelectric crystal to replace the mechanical spring in an alarm clock. Let's look at simple springs first, piezoelectric crystals second.

Springs make for very good timekeepers. Springs resist being stretched or compressed, and respond with a force that opposes the change in their length. The more the spring is stretched or compressed, the greater the opposing force. Hang a coiled spring from the ceiling, and attach a weight to its end. The spring will stretch down, with the spring providing an upward force, opposing the stretching and balancing the downward gravitational pull of the weight. If you pull the weight down a bit more and let it go, the upward force from the spring is now larger than the downward force of the weight, and the weight moves upward, speeding past its original position. As the weight overshoots, it compresses the spring, and the coil responds with a downward force, now opposing the squeezing and pushing the weight back toward its starting location. The weight will go up and down periodically, with a motion no different from the swinging pendulum bob and the alternating voltage from the wall outlet. The natural frequency of the spring's oscillations (the number of up-and-down cycles per second) is determined by its stiffness and how much weight is hanging on its end.

The origin of the force in a spring that resists being stretched or compressed is the same as in the piezoelectric crystal in your smartphone—electricity. The atoms in all solids are held together by electric forces, which also make sure that the atoms stay in specific locations. If two neighboring atoms in a solid get too close to each other, there is a repulsive force between the electrons from each atom that pushes them back apart. Imagine an atom in a crystal as a simple ball. Let's represent the chemical bonds holding the atom in place as springs attached to the atoms on either side. Push this atom away from its natural position in a solid, and its surrounding electrons will encroach too closely on its neighbors on one side, and be too far away on the other. This creates an unbalanced force that will push the atom back toward its equilibrium position. That force will decrease by the time the atom is back to its natural spot in the crystal, but due to its kinetic energy it will overshoot and now move toward the neighboring atom on the other side. It will oscillate back and forth around its preferred location, with the amplitude of vibration depending on the solid's temperature, and the frequency of oscillation depending on the mass of the atom and the stiffness of the chemical bonds holding it in place in the solid. This vibration of the atoms happens in *all* solids: this book, the chair you are sitting in, even you yourself.

Electronic timekeepers, such as a digital wristwatch and your smartphone, use a special oscillator that is much more accurate than a coiled spring—a quartz crystal. Quartz is a solid composed of molecular units of silicon dioxide, the chemical composition of sand. Quartz crystals have a special property: the electric charges of the molecular units line up to create a net electric field along the length of the solid when it is squeezed in one direction. This type of material is called piezoelectric: *piezo*, in Greek, means "to squeeze or press," and a piezoelectric material is a solid that generates a voltage when it is squeezed. For certain materials and crystal structures, when two sides of the solid are pushed together, all the atoms buckle in just the right way to create a large, net electric field.

To use a piezoelectric crystal as a timing device, we run this process backward. That is, we apply a voltage across the solid, and the crystal sides pull together, as if compressed by an external force. Once the voltage is turned off, the crystal will expand, and it begins to oscillate at its natural frequency. That frequency is determined by the size and shape of the crystal, and can range from a few thousand cycles per second to as high as several hundred million cycles per second. As the quartz crystal oscillates, it generates a voltage at this same natural frequency that can be used to maintain the crystal's vibrations. As in the digital timer, computer chips reduce the quartz crystal's high frequency down to one cycle per second. Once the preset time is reached, a signal voltage is sent to another chip. In the coffeemaker, this second chip starts the brewing process for the coffee, while in your phone it initiates playing a preselected musical tune.

Timing is very important for a smartphone, aside from any alarm clock functions. Any procedure by a computer (and for our purposes, the smartphone can be considered a small computer) is an operation that exists in time. There is a start to the process, and an end—not unlike a piece of music. Moreover—again, not unlike a piece of music—in order to achieve the desired effect, the notes have to play in the appropriate sequence, at the right time. For a symphony orchestra to be effective, the musicians need a good conductor who will stay on beat and ensure that, for example, the horns don't begin playing the fourth movement while the strings are tuning up. In a computer, with millions of transistors, logic elements, and memory cells that all have to perform in the right sequence, the "musical conductor" is a chip called the central processing unit, or CPU. With the help of a crystal oscillator, the CPU is able to keep the beat—and a very fast beat at that—in order to coordinate elements that switch in less than a nanosecond.

After freshening up, you grab your smartphone from the bedside table and head to the kitchen. You take a bagel and a stick of butter out of the refrigerator, and leave them on the counter to warm up. Finding a podcast you have already downloaded on your phone, you listen as you begin getting ready for the day. This particular podcast features profiles of renowned makers of violins, contemporaries of Stradivarius. The program intersperses the biographies of these woodworking artisans with examples of classical music played on their magnificent instruments. In order to better appreciate *the snippets of music, you connect your smartphone to two small but rather high-quality **speakers** that sit on your countertop.*

To play your podcast, or any music stored in its memory, your smartphone has to convert a numerical code into sound waves, which are changes in the density (hence pressure) of the air. The ability to store music predates the electronic era, with simple windup music boxes able to play a brief snippet of a single tune, and player pianos capable of performing the entire number. The methods by which they store their information differ greatly, but one thing the music box, player piano, and smartphone all share is that eventually, in order for the sound to be heard, there has to be a vibration in the air.

An MP3 player[*] uses a digital instruction set that mimics how the holes work in a player piano roll. A player piano has an internal mechanism that can autonomously instruct which keys to depress, via a roll of paper with holes strategically arranged. Variations on the holes' locations and spacing in this long sheet basically hold a code that, when interpreted by the piano's mechanism, will play a particular song. The digital in-

---

[*] This notation is a shorthand to describe the software used to compress the music so that fewer ones and zeros are needed for the instructions sent to the speakers.

formation stored in your smartphone is also essentially a code that, when read properly, creates a pattern of voltages. When sent to a speaker, these voltages are converted into the sound waves of the specific musical tune. Inside the speaker is a membrane (a very thin plastic sheet) that can vibrate. Depending on the frequencies and amplitudes of the membrane's vibration, pressure waves are generated in the air—and it's these waves that we can hear.

How does one convert electrical voltages into mechanical vibrations of the membrane so that we can hear the resulting sound waves? It's done with magnets. Attached to the membrane is a small coil of wire. A voltage is used to drive a current through the coil, and any changes in the voltage are reflected in the current. These current changes are converted to mechanical vibrations on the speaker's membrane using the same symmetry between electrical currents and magnetic fields encountered before. The varying electrical currents generate varying magnetic fields. The coil on the membrane sits right on top of a permanent magnet directly underneath the diaphragm. When the current flows in a clockwise direction, it generates a magnetic field oriented so that the north pole faces outward, toward the north pole of the permanent magnet. Because identical polarities repel each other, there is a force pushing the magnets apart, causing the membrane to flex outward. When the voltage direction is reversed, the current flows in the opposite direction (counterclockwise), and the magnetic field generated has a south pole facing the permanent magnet's north pole. Opposite magnetic polarities attract, and the coil is pulled toward the permanent magnet, causing the membrane to distend inward. Changes in the voltage, in both frequency and amplitude, cause back-and-forth modulations in the membrane, which in turn generate sound waves.

An earbud places the vibrating membrane in proximity to the eardrum. For a conventional stereo system, the speaker's membrane is at the center of a larger cone that flares out, amplifying and projecting the membrane's vibrations. A smartphone's speakers are in its case, so the quality and volume of the music it plays are compromised. (If you want a quick and not-very-dirty way to amplify the sound from your smartphone speakers, rest the phone's speaker on the bottom of a large bowl, preferably one made of wood, and the sound will have a richer and deeper tone. The natural frequencies of wood enhance the reflected sound waves—making it the material of choice for string instruments.)

You get caught up in the podcast as it plays a vibrant segment of music. You imagine the virtuoso rhythmically plying the bow back and forth across the strings. The image of the bow suddenly reminds you that you still need to brush your teeth. Turning up the volume, you dash to the bathroom and

*remove your **electric toothbrush** from the plastic holder
that serves as the battery recharger.*

To make the bristles move back and forth automatically, your electric toothbrush employs a small electrical motor, connected to a rechargeable battery. The conversion of electrical potential energy stored in the battery into rotational kinetic energy underlies a great deal of the technology we encounter daily. The battery causes a current to flow in a wire coil in the motor. The coil is situated between the north and south poles of a small magnet. The current in the wire creates a magnetic field that is repelled from one pole of the magnet and attracted to the other pole, twisting the coil. The coil is attached to a rod passing through its diameter, and when it twists, the rod rotates. By using a clever switching mechanism, the battery's direct current (DC, as opposed to the AC from the wall outlet) can be made to change direction every half-turn, so that the coil is continually being repelled by one pole of the fixed magnet and attracted to the other pole. Through an off-center rod, the toothbrush's internal motor's rotational motion is converted to a back-and-forth motion—causing the toothbrush bristles to oscillate against your teeth.

The motor in some electric toothbrushes spins at a rate of several hundred revolutions per second, and this is the source of the hum you hear when cleaning your teeth. Some toothbrushes can have frequencies as high as 1.6 million cycles per second (but with very low amplitude oscillations). These high-frequency vibrations are created using frequency multipliers (just as we saw in the digital timer) and a piezoelectric crystal (found in the timer in your smartphone).

The source of electrical energy in your toothbrush is a battery that uses a chemical reaction to pile positive charges up on a metal bar (called an electrode) and negative charges on a different electrode. This charge difference drives the current through the coil in the motor (or gets the piezoelectric quartz crystal oscillating). The chemical reaction eventually reaches a point where no more charges can be added to the electrodes. In rechargeable batteries a voltage is applied across the terminals, forcing the chemical reaction to run in the opposite direction. With the battery reset, the forward chemical reaction can again continuously charge up the electrodes, and the battery is once more ready for action.

But how to electrically connect the batteries to a power source in a device that, by its very design, will be near or submerged in water? Many electric toothbrushes have a plastic handle, which is placed in a cylindrical recharger that—while connected to an electrical wall outlet—is also solid plastic. Plastic is an electrical insulator; the electrons in each atom are devoted to holding the atoms together, and no electrons are available to carry a current in response to an external voltage. So how does a plastic-to-plastic connection from the recharger to the handle of the toothbrush supply electrical energy to the recharge-able battery? Through the same physics that describes how changing magnetic fields create electrical currents.

In the base of the toothbrush's recharger is a wire coil. When the recharger is connected to the wall outlet, an alternating current flows through the coil. This current continually reverses itself, flowing clockwise and then counterclockwise like the back-and-forth motion of a pendulum. The magnetic field generated by this current will also continually change direction, first with a north pole facing out from the bottom of the charger and then, in the next half-cycle, a south pole facing outward. The base of the electric toothbrush contains another coil, oriented so that the magnetic field of the recharger (or most if it) will pass through the circular area of this loop of wire. The strength and the orientation of the magnetic field passing through this second coil are always changing, thereby inducing a current through the same mechanism used to generate electrical power at a power plant. This current in the toothbrush's coil, set in motion by the process of magnetic induction, is converted from an alternating current to a direct current, and then used to recharge the battery in the appliance. A device in which the current in one coil induces a current in a second coil, even though the coils are not directly connected, is called a "transformer" and has many uses—not just recharging the battery in your electric toothbrush.

If the coil in the recharger has the same number of turns (like the winding of thread on a spool) and the same diameter as the coil in the base of the electric toothbrush, then the current induced in the toothbrush coil will be the same as the current that flows in the first coil (assuming that all of the magnetic field from the recharger coil passes through the coil in the toothbrush base). But if the second coil has more or fewer turns, then the current induced will be smaller or larger, respectively, than in the first coil.* This is advantageous, as the alternating current coming from the wall outlet has a peak value of 110 Volts, which is too large to send to the battery in the toothbrush. Transformers are used all the time either to boost the voltage in power lines to improve the transmission efficiency or to decrease the voltage, when the power line reaches your house, to a safer, lower voltage of 120 Volts, suitable for the appliances on your kitchen countertop.

*Back in the kitchen, you slice your bagel into two halves. You place each half into the **toaster**, and depress the lever. The spring inside the toaster that holds the slices up is com-*

---

* Electrical power is mathematically described as the product of the current and the voltage, and a larger current in the second coil means that the voltage will be smaller. (One is constrained, after all, by conservation of energy, and one can't get more power out of the system than one puts in.)

pressed, and the bagel withdraws into the toaster, where the wires begin to warm, eventually glowing red. The butter is still a bit hard, so you put it in the butter dish and place the dish atop the toaster's openings; the heat will make the butter softer and easier to spread.

The toaster is a technology that would be familiar to your great-grandparents. When you put a piece of bread in a toaster and push the lever arm down, in addition to lowering the slice into the toaster, you are also closing a circuit that allows an electrical current to flow through the wires adjacent to the bread. After half a minute or so, the wires become warm and then begin to glow red-hot. Why? To understand how the toaster converts electrical energy into heat and light requires an understanding of thermodynamics, electromagnetism, and quantum mechanics. All for a piece of toast!

A toaster employs the first law of thermodynamics, which states that for any closed system, the total amount of work and heat must remain unchanged. When you close the circuit by pressing down on the lever, the current is forced through the wire, and thanks to the resistance in the wire, the electrical current's energy is converted into heat.

Let's start with the metal wire itself. To be a good

conductor of electricity, a material needs a large number of electrical charges that are free to move. Metals have a large density of mobile electrons and thus make excellent carriers of electrical current, while insulators like plastic or glass have their electrons tied up in the chemical bonds between atoms. The arrangement and detailed chemical interactions of atoms in a solid, subject to quantum mechanical constraints, determine whether any given material is an insulator or a conductor.

Toaster wire is usually composed of an alloy of nickel and chromium (called nichrome), both metals that can carry an electrical current. For quicker toast, the wire should be a good conductor of electricity—but not too good. It's the mixing of two different metals in the nichrome wire, along with any other defects or imperfections, that will lead to the desired heating.

Think of the wire in a toaster as a large staircase with a throng of people all trying to descend at the same time. The more people exiting at the bottom of the stairs, and the faster they are moving, the greater the current. The voltage, which here would be the steepness of the stairs, is what gets people moving down in the first place. A very steep angle would translate to a large voltage, meaning a single person would exit the bottom of the staircase with a greater speed. The individual steps correspond to the atoms in the metal. It's easier to descend the staircase, especially when you are part of a large group, by having everyone line up across the width of the staircase and move down, step by step, in unison. As one row exits at the bottom, another starts off at the top—allowing the staircase to achieve its peak efficiency.

However, people in a staircase, like electrons in a wire, move somewhat randomly rather than marching in lockstep. Moreover, real staircases (and real wires) are not so uniform. If there is a step missing (in a wire this might be an atom out of place, for example) that no one notices until they try to walk onto it, there would be a large tumble. All of this leads to inefficiency, and the whole journey takes longer—which (in toaster-wire terms) translates to a lower current.

In a metal, this is characterized as contributing to the wire's "resistance." There are geometric contributions to resistance, as it's harder to pass a current through long and skinny wires compared to short and fat ones. In some applications, resistance is a significant hindrance. But when making breakfast, it comes in handy.

The resistance of the wire leads to a transfer of kinetic energy from the electrical current to the atoms in the wire, causing them to vibrate more violently than before, a process known as "Joule heating."* This is why nichrome is used in your toaster—it's a good enough conductor to carry a current, but it also has a large

---

* This same physics applies in space heaters and hair dryers.

resistance, to maximize the Joule heating. With enough transferred kinetic energy, the atoms near the defect can shake so violently that they emit light, which is why the toaster wire glows.

When an atom vibrates, the electrons swing back and forth like a mass on a spring, forming an electrical current that fluctuates inside the atom. That electrical current generates a magnetic field. As the current is continually alternating in magnitude and direction, the magnetic field is constantly changing, and the changing magnetic fields generate electric fields. Think of the charged pendulum bob shown on page 5. The periodically varying electric and magnetic fields combine to form an oscillating electromagnetic wave called "light."

To create that perfectly done piece of toast, the heat of the toaster wire (which can be over 1,000°F) is transferred to the bread. In hair dryers and space heaters, air molecules are pushed past the hot wire, picking up excess kinetic energy, while in the close confines of a toaster, heating occurs mostly from infrared radiation. When the surface of the bread is approximately 300°F, sugars and starches undergo a chemical reaction, becoming brown and changing their flavor and texture. The "toast setting" knob is actually an adjustable resistor that varies the current in the toaster wires. A timer or temperature sensor is used to open the electrical circuit and stop this process, hopefully before your toast burns.

*You remove the butter dish from the top of the toaster, just before the butter melts. A few moments later the bagel pops up, toasted to perfection. As tempting as it is to relax with your cup of coffee, you have a doctor's visit to get to. You pour yourself a glass of orange juice and listen to your podcast while you finish your breakfast. The calendar reminder on your phone buzzes to alert you that your doctor's appointment is in an hour. You clean off your plate and rinse out the juice glass and coffee mug, placing them all in the dishwasher, and return the orange juice bottle and butter to the **refrigerator**.*

A toaster makes use of the first law of thermodynamics, in that the work done forcing a current through a wire is converted into heat. The second law of thermodynamics places limits on how well we can run this process in reverse, extracting heat and using it to do work, as in a refrigerator.

The development of the field of thermodynamics subverts the standard paradigm, in which the basic science research comes first and the practical applications follow. Rather, steam engines—a practical application of a little-understood science—came first, and only later, motivated by the desire to improve the engine's efficiencies, did scientists figure out the basic physics. An engine converts random heat into useful work, as in the burning of gasoline in an internal combustion engine, while a refrigerator is an engine run backward, where work is done to remove heat from a system.

A refrigerator lowers its internal temperature using the same physics you rely on to cool off a hot cup of coffee by blowing on it—evaporation cooling. Say your morning cup of coffee is too hot to drink. Temperature is a measure of the average kinetic energy of the molecules in the coffee. That is, some molecules will have a kinetic energy below the average, and some will be much more energetic. Those more energetic, eager-beaver molecules form the cloud of steam over your coffee cup and have enough kinetic energy to initiate a phase transition, moving from the liquid state into the vapor phase. When you blow on your coffee, you are pushing these high-kinetic-energy molecules away from the cup, preventing them from returning to the liquid and redepositing their energy into the liquid. With those high-energy molecules no longer part of the coffee liquid-vapor system, the new average kinetic energy of all the molecules is lower than it was before, reflected in a lower temperature for your coffee.

Your refrigerator basically operates on the same principle but uses a different liquid instead of coffee. Refrigerators once used Freon but now have transitioned to tetrafluoroethane.* An electric motor is used to run

---

* In 2016, 170 nations agreed to reduce the use of hydrofluorocarbons such as tetrafluoroethane in refrigerators and air conditioners. Alternative coolant fluids can be used without changing the basic physics by which a refrigerator operates.

a mechanical pump, which forces the coolant liquid through a thin metal pipe. Metal is a good conductor of heat (the "sea of free electrons" can carry energy as well as electrical current), and it ensures a good thermal connection between the coolant liquid and the walls of the fridge. The pump forces the coolant through an expansion valve, allowing it to go from the narrow tube into a larger volume, where it undergoes a phase transition from the liquid to the vapor state. You have to add energy to a liquid to convert it to a vapor (consider boiling water), and that energy has to come from somewhere.* The cloud of coffee vapor extracts kinetic energy from the rest of the liquid coffee in order to evaporate, while the tetrafluoroethane pulls heat away from the interior walls of the refrigerator. The coolant liquid moves through a tube that makes a series of S-shaped turns in order to maximize its surface area in contact with the fridge walls. The density of S-shaped turns is higher in the freezer section so that more heat is extracted from this volume.

What to do with the energetic gas once it has removed heat from the refrigerator? To repeat this process and keep the fridge permanently cool, the pump is now used to recompress the gas back into a liquid. When the

---

* This is also the mechanism by which we cool ourselves when we perspire. Sweat only cools when it evaporates, extracting energy from the skin to move from the liquid to the vapor state, lowering the average kinetic energy of the body. If the atmosphere is saturated with water vapor on a muggy day, then this process is inhibited and we can't cool off as effectively. So physics says it really is true—it's not the heat, but the humidity!

vapor transitions into the liquid, it gives back the heat it extracted when it went from the liquid to the vapor. Because it takes energy to run the pump, there is a net cost of energy in running the refrigerator. The tubes in this part of the closed-cycle system are placed behind the back of the fridge, next to the wall, so that they do not return their heat to the interior of the refrigerator. When a temperature sensor indicates that the desired internal temperature has been reached, the pump is turned off. There are thermal leaks at the seals of the doors to the fridge and freezer, though better engineering and better materials have improved the quality of these seals. If you leave the door of the fridge open while staring at the interior contents, contemplating some deep truth of the universe, you will also cause the interior of the refrigerator to warm. Then you will hear the compressor of the fridge turn back on.

CHAPTER TWO

You Drive into the City

*You take the elevator down to the garage of your building. You head over to your **electric/gasoline hybrid vehicle** and unlock your car door using the keyless remote control on your key fob. When you are settled behind the steering wheel, you start the car and drive to the exit. You slow down for a moment, and a sensor registers the electronic key in your car and opens the garage doors.*

The automobile is essentially a machine for the conversion of potential energy into kinetic energy. In an internal combustion engine, the potential energy is chemical, stored in the gasoline molecules; in an all-electric car, the potential energy is in an electrochemical battery. A hybrid car uses both an internal combustion engine and an electric motor, maximizing the advantages and minimizing the limitations of each process. Vehicles with gas engines can travel great distances on a single fuel tank, but their mileage is generally poor and they produce noxious exhausts. Electric cars are clean and efficient, but they have a limited driving range owing to the relatively poor energy density of batteries. For driving under normal conditions, a much smaller (and hence more fuel-efficient) engine would suffice. However, this smaller engine would not be able to provide sufficient power for accelerating to highway speeds or climbing a steep hill. In order for a smaller, more efficient engine to provide the necessary thrust in those less frequent but crucial situations, a hybrid automobile calls upon a battery-powered electric motor as a secondary energy supply. The net result is an automobile that has much better mileage due to its more efficient engine, without sacrificing acceleration thanks to its electric motor assist.

But it does not matter whether the energy that rotates the tires comes from a battery, a gasoline-fueled internal combustion engine, or even an old-fashioned steam-driven antique—from a physicist's perspective, every single automobile is an all-electric vehicle.

Normally, atoms are electrically neutral (they have no positive or negative charge). Electrically neutral atoms will not form chemical bonds with other atoms—unless the conditions are just right. Bring two atoms very close together, and their electron orbits will overlap. The negatively charged electrons will repel each other, keeping the two atoms separate and distinct. In order to overcome this repulsion, whenever two atoms link up and form a simple molecule, there must be an additional attractive interaction that overwhelms the repelling electrons. All bonds between atoms depend on electrical attractions to hold the atoms together. The bonds between atoms in a molecule of gasoline, called "covalent bonds," are generally very strong; it takes a fair amount of energy to break them. When these molecular fragments re-form, as in the combustion reaction in an automobile engine, they can release a similarly large amount of energy.

Another type of chemical bond, termed "ionic," is responsible for the energy stored in a battery. Atoms that have fewer or more electrons than normal are called "ions"; a "positive ion" has fewer negatively charged orbiting electrons than positively charged protons in its nucleus, and one with more electrons than protons is a "negative ion." Batteries are devices that use ions to create a voltage. A typical battery employs ions moving between two metal rods, the "electrodes" or "terminals," which typically reside in a fluid, either acidic (such as sulfuric acid) or alkaline (typically potassium hydroxide),* that can strip charges from the atoms in one terminal, promoting the formation of ions. By choosing the right metals for the electrodes, and the right chemical fluid, we can induce negatively charged ions to pile up on one rod and positively charged ions to pile up on the other. A separating barrier in the fluid keeps these ions on the terminals and prevents them from recombining and discharging through the fluid. When a wire is connected across the terminals, this charge difference pushes the electrons in the wire away from the negative ions on the one terminal and toward the positive ions on the other. The net result is that an electrical current flows through the wire. This electrical current can be used to do mechanical work, such as powering an electrical motor.

When the battery is used to drive a current in an electrical circuit, the ions stored on the electrodes are effectively removed, and the chemical reactions in the battery fluid can then continue to add negative and positive charges to the metal rods. But eventually the reactive components of the fluid are used up, the battery cannot maintain its specified voltage, and the battery is "dead." Fortunately, the batteries in today's all-electric vehicles are rechargeable.

---

* Elements from the first column of the Periodic Table of Elements, such as lithium, sodium, and potassium, are termed "alkali metals"; alkaline batteries use such elements to facilitate the transfer of ions.

Another limitation on a battery's stored electrical energy is that the metal terminals can hold only so many charged ions, a major drawback when using batteries as a sole source of automotive power. Compared to the energy density of a battery, the energy density stored in the chemical bonds in a molecule of gasoline can be much higher. The trick is extracting that energy in a way that leads to tire rotation.

A standard automobile internal combustion engine generally consists of four to eight cylinders, in which different chemical vapors are injected through a hose attached to a hole near the top of each cylinder. The cylinder has rigid side-walls and a fixed top, and a bottom plate (called the "piston") that can slide up and down.* There are four steps in a combustion cycle, with the first two being the injection of gasoline vapor and oxygen into the cylinder, followed by the movable piston being forced up, compressing this gas mixture. The gasoline and oxygen molecules, squeezed into a smaller volume are now hotter, and have greater kinetic energy. This hotter gas is just below the temperature at which the gasoline molecules will react with oxygen and undergo combustion. Step three begins with the input of trigger energy, in the form of an electrical arc from a spark plug

---

* In some configurations the bottom is the fixed cap and the top is the movable piston, but this changes nothing in the physics of the four-stroke cycle.

that ignites the hot gasoline vapor. Chemical bonds in the gasoline molecule are broken and new bonds form, with the net result that the chemical products have far greater kinetic energy than they did before the spark. These faster molecules strike the inner surface of the cylinder, exerting a large pressure and forcing the piston down. Stage four: the gases (the ignited air/gas mixture and any unreacted gas in the cylinder) are pushed out as exhaust through another hose, returning the cylinder to its initial position.

Internal combustion engines repeat this four-step cycle as we drive our cars. The up-and-down motion of the piston is converted to rotational motion of the wheels through a clever mechanical coupling, not unlike that employed in an electric toothbrush. In the automobile, a rod attached to the top of the piston is attached to the edge of a disc. As the rod moves up and down, it causes the disc to spin, and the rotation is transferred to the tires.

Why does burning gasoline release energy? A gasoline molecule consists of a chemically connected chain of covalently bonded carbon atoms (typically seven to eleven), strung together like beads on a bracelet, with hydrogen atoms chemically bonded to each carbon atom along the chain. A molecule is stable if its collection of bonded atoms has a lower energy than when they are physically separated and distinct. This lower binding energy, which results from quantum mechanical inter-actions between the electrons in each atom, holds the co-valently bonded molecule together. When energy greater than its binding energy is supplied to the molecule, it may "unbind" and break into its constituent pieces; the molecular fragments can then react with other atoms, forming new molecules. These molecules will be lower in energy than the separate atomic building blocks, and as they form, moving into lower-potential-energy config-urations, conservation of energy will speed up the chemi-cal products (that is, they will have more kinetic energy). This excess kinetic energy of the reaction products is re-ferred to as "heat." Different molecules will release dif-ferent amounts of kinetic energy through combustion. One reason that the four-cycle internal combustion en-gine ultimately won out over steam- and electric-powered automobiles at the beginning of the twentieth century is that gasoline has, pound per pound, one of the highest energy densities available.

In some hybrids, the electric motor and the gas-fueled engine operate simultaneously to provide kinetic energy to the tires. Other hybrids use an either/or configuration, with the electric motor starting the car and the inter-nal combustion engine taking over when it's at cruising speed. Either the gas-fueled engine runs a generator that recharges the battery, or the battery is recharged every time the car brakes. In the latter case, the rotational kinetic energy of the tires is transferred to a generator that maintains the battery's charge. Either way, the battery never needs to be externally recharged.

*You begin driving down familiar streets and are about to merge onto the highway you routinely take to work, when that coffee finally kicks in and you remember that you're not going into the office today. You have accidentally driven in the opposite direction of where you want to go, away from the doctor's office. You are running late, so you consult the **global positioning system** built into the dashboard of your car. You quickly type in your doctor's address, and in no time the GPS provides step-by-step instructions to your destination. You pass a series of gas stations before the entrance ramp of the expressway, and appreciate that you no longer depend on gas station attendants for directions.*

Your GPS device communicates with satellites to determine your exact location. These particular satellites are in medium Earth orbit, roughly 12,600 miles above the Earth's surface, and take approximately twelve hours to go around the planet. There are thirty-two such satellites orbiting Earth,* and no matter where

---

* This refers to the U.S. system, which is shared with other countries. However, in order not to rely on America, China has recently deployed its own network of positioning satellites.

you are, you are within radio contact with at least four of them. These satellites broadcast radio wave signals at periodic intervals that indicate the time and their location to any device that has a functioning receiver, such as your GPS. Your GPS device detects these signals and calculates how long it took for each of them to arrive, based on when the satellite sent the message. Knowing the transit time and the speed of the signal (radio waves are just another form of light, so this speed is the speed of light), your GPS device checks its position relative to the satellite, and to the signal received from two other satellite signals. Comparing the three different transit times to the location relative to three orbiting satellites, it is able to accurately pinpoint your location.* Once you know where you are, programmed maps stored in the device's memory instruct you how to get where you want to go.

The better the timing of the signals, the more accurate the determination of your position. The speed of light is so fast that even small errors in the timing can yield large errors in the positioning. An object, such as the satellite, that is far away from a large gravitational mass (the Earth) will have its clock run a little faster

---

* Each satellite in essence generates a circle, and your receiving device can be anywhere on its circumference. The three circles from the three satellites overlap and cross, and where they all intersect will be your location (the only point where you can be on the circumference of the three circles simultaneously).

than one on the large mass's surface. To understand why gravity will affect the satellite's clock speed, we must turn to Albert Einstein's general theory of relativity, which is a theory of gravity. Whether you use an onboard navigation system or the locator on your smartphone, your GPS is accurate thanks to Einstein's theoretical genius.

You are familiar with the effects of gravity; you encounter them when you stand on a bathroom scale. The recorded weight is due to the scale pushing up to balance the downward gravitational pull of the Earth on your body. If you were in a closed room in outer space, you would be weightless. But if the room were to be accelerated by an external force (say, by a cable attached to the ceiling), then you would no longer float in the room; the floor would instead push up on you. This acceleration from the cable would thus cause the scale to display a number. Einstein recognized that there is no way to determine whether the scale's reading is due to gravity or due to the cable providing acceleration, and that therefore the two must be equivalent.

Gravity results from interactions between masses, and acceleration describes how we move through space and time. Mass distorts the geometry of space and time, and this distortion changes our trajectories.

The Earth has a large mass, and it warps space and time around it. Think of a thin trampoline sheet, held taut at its edges, onto which one places a bowling ball.

The ball sags down, and the sheet bends around the ball. A smaller mass (say, an orbiting GPS satellite) moving near the bowling ball will change its motion as it follows the curvature of the sheet. If the speed of the smaller ball is chosen correctly, it will follow a circular orbit around the bowling ball. This orbit can be thought of as a consequence of moving on a curved surface, or due to a gravitational pull, and Einstein's theory says that they are the same thing. To paraphrase theoretical physicist John Archibald Wheeler, in general relativity, matter tells space how to bend, and space tells matter how to move.

There are *two* effects stemming from relativity that can throw off the timing of the highly accurate atomic clocks on the GPS satellites. The satellites orbiting the Earth cover a distance of over 104,000 miles in only twelve hours and are traveling at a high velocity relative to you on the surface. The clocks on the satellites will thus run slower than clocks on the Earth's surface—a time dilation effect described by Einstein's special theory of relativity.

And clocks run slower the closer they are to a gravitational mass. To see why, think about space curved by the Earth's mass. The distance that light will travel between two points in a given time period is stretched out when it is closer to the Earth's surface. But if light is to cover this longer distance with the same speed (a central principle of relativity is that the speed of light in a vacuum is the same everywhere) then time must slow down closer to the Earth's surface. That is, the greater the gravitational pull on a clock, the slower it will run. Knowing the gravitational pull of Earth, one can determine how much faster the clock on the satellite is running compared to an identical clock on Earth. The two effects—the satellite's clock running slower due to special relativity and it running faster through general relativity—are not equal and do not fully cancel each other out. The net change is a faster clock on the satellite from the gravitational effect. If one did not take these small but measureable time shifts into account and correct for them, GPS systems would be of dubious value.

*You pull onto the expressway, this time heading in the right direction. You smoothly accelerate down the entrance ramp and merge into a middle lane. The GPS device on the dashboard indicates that you need to continue on this road for another four miles, which means that you will have to pay the highway toll. As you approach the row of tollbooths, your heart momentarily sinks when you see a series of brake lights ahead. You are relieved that the slowing down is only for the two far-right lanes, which are for those drivers who must pay cash to enter the highway. You glance at the small white box on your windshield, grateful for the **E-ZPass** device that automatically pays your toll. You cruise past an open* *tollbooth, able to completely bypass the bottleneck of cars whose drivers are paying their tolls with legal tender.*

The E-ZPass system—along with garage door openers, keyless remote entry devices, and walkie-talkies used by police, the military, and some emergency workers—basically relies on specialized radios. One part is a small battery-powered receiver, like the E-ZPass device in your car, while the analog of the radio antenna is in the tollbooth. Not unlike a walkie-talkie set, both the car's receiver and the tollbooth antenna can broadcast and receive instructions sent via radio waves. Some E-ZPass systems initiate the receiver/ antenna conversation when the automobile intercepts a light beam that spans the width of the tollbooth. When the lane is empty, that light beam strikes a photodetector. But when a car blocks this light by driving through the tollbooth, the absence of a signal at the photodetector breaks a circuit. This leads to the closing of another circuit, causing the antenna to send out a signal to the receiver in the E-ZPass device, beginning a set of radio transmissions between the tollbooth and your car.

If E-ZPass systems, garage door openers, and walkie-talkies are all basically specialized radios, then you might ask: How does a radio work? The physics behind the generation of radio waves is the same science that accounts for the toaster wire's red glow. As the electrons in the toaster wire's atoms shake back and forth due to its rising temperature, they create electromagnetic waves (since changing electric currents create varying magnetic fields and, in turn, changing magnetic fields generate electrical currents). The toaster wire emits both infrared and red light. Electrons oscillating at much lower frequencies create radio waves.

Physicists use the word "light" to describe any electromagnetic wave. Radio waves, microwaves, infrared light, visible light, ultraviolet light, x-rays, and gamma rays are all light, differing only in the frequencies of their oscillating electric and magnetic fields. The E-ZPass system uses radio waves with a frequency of 900 MegaHertz (that is, 900 million cycles per second).*

When the radio waves emanating from the tollbooth's antenna encounter electric charges in a wire in the E-ZPass device, the oscillating electric field will cause the electrons to vibrate back and forth. In the E-ZPass system the tollbooth and your device both receive and broadcast signals, so which one is called the "receiver" and which one the "antenna" is arbitrary.

There are many ways to use these waves to encode information, but two have dominated: varying the peak heights of the wave, called "amplitude modulation"

---

* This is the same frequency that many cordless landline phones use. But there is no risk of interference with your phone at home, as the E-ZPass system is low-power and thus has a very short range.

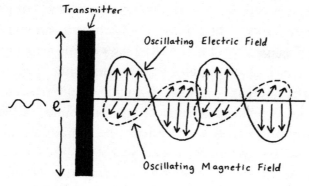

Transmitter

Oscillating Electric Field

Oscillating Magnetic Field

Changing Current
into Transmitter

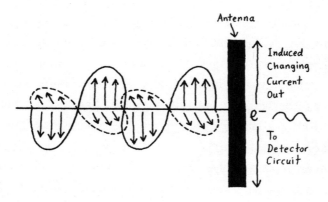

Antenna

Induced
Changing
Current
Out

To
Detector
Circuit

When the electromagnetic wave strikes the antenna,
it causes electrons to oscillate up and down, creating
a CHANGING CURRENT.

FIGURE 2

(AM), or changing the oscillation frequency, termed "frequency modulation" (FM). Both require modulations of electric currents more involved than having a charged mass oscillating back and forth on a spring. The detected electromagnetic wave in a radio is converted to a voltage that makes the receiver's speakers vibrate so that the resulting sound waves mirror the initial message. The E-ZPass system does away with the speakers, but the basic process is the same. Information is encoded in the variations of the radio waves, and there is a back-and-forth communication between the tollbooth and the device in your car.

Upon receipt of this radio wave signal, the receiver in the E-ZPass device broadcasts a different radio wave signal, identifying the car's account information. This digital signal is then sent to a small computer that verifies the account and checks that there are sufficient funds available for the toll. Another message is sent to a display informing the driver that his or her account has been appropriately charged. Radio waves, traveling at the speed of light—186,000 miles per second*—accomplish all this before the automobile has progressed even a fraction of an inch through the tollbooth. While the photodetector records the passage of an automobile, the

---

* This is the speed in a vacuum. The speed of light can be slower in other media, such as water or glass—which we'll return to later on.

tollbooth sends out a signal looking for the E-ZPass. If no responding signal is received, then another signal is sent to a camera that photographs the car's license plate. This will eventually lead to a fine (sometimes delivered by the much-slower-than-light-speed postal service).

*Traffic seems to be going at a decent pace, and you begin to wonder if you might actually arrive early to your appointment. Spotting some gaps in the traffic flow, you switch lanes twice, trying to advance and make better time. You glance away from the road for an instant and turn on the radio to check the weather and traffic reports. A nice, partly cloudy day with no significant storms or wind—looks like there won't be any problems with your flight departure. The traffic report is equally positive, with no reports of any significant road construction or automobile accidents during the morning rush hour. Suddenly you see a series of brake lights ahead. You quickly step on the brakes to avoid rear-ending the car in front of you. There is an actual **traffic jam** that seems to have popped out of nowhere. You smack the steering column in frustration, wishing you could levitate over the stopped cars.*

As the density of cars per mile of highway increases, the "flux"—that is, the number of cars passing a given point per hour—will grow. But if the density of cars becomes too large, the flux actually decreases, and a lower flux translates to a longer travel time. Physics says that traffic would be forever smooth and easy, if only we could get rid of the drivers.

When the density of cars on the highway is low, the motion of each one can be treated independently, its speed set by road conditions but not constrained by other vehicles. This is not unlike a very dilute gas of atoms, where the chance of two atoms meeting each other is small. The situation changes as the density of cars increases, when the motion of cars is more like the atoms in a liquid, where movement is limited by their interactions with neighboring atoms. At rush hour, when the density of cars on the road is highest, you can become stuck in a traffic jam that seems to have no obvious cause. These jams are not always caused by construction or an accident; they are, in fact, an intrinsic instability of the traffic itself.

For higher car densities, the flow of traffic can be described as a collective phenomenon, not unlike when water molecules interact with other nearby molecules to form a large-scale disturbance, such as a wave. In the case of highway traffic, when drivers spot a gap in traffic ahead, they typically do not slow down to increase the spacing between their car and those in front of them, but rather speed up to decrease this separation. Each driver deliberately, though independently, packs his or her car closer to those ahead, forming a cluster. As faster cars meet the rear edge of the cluster, they must slow down; they can speed up again only when they move through the high-density pack and reach the cluster's front edge.

Two factors play a key role in smooth traffic flow: your awareness of the buildup of cars ahead of you, and your response time to any change in the density of cars. A long response time means that it takes longer for drivers to respond to changes in density in front of them, and instead of smooth flow, the clusters will grow larger and larger.

An analogous situation is a sandpile that is built up by dropping dry grains of sand on top of each other, until an unstable cone is formed. If one more grain is added, it can cause an avalanche down the side of the pile. Similarly, at rush hour, cars pack themselves into ever-growing clusters, and then all that is needed for a jam to form is one person at the front edge of the pack tapping on the brakes or just easing off the gas pedal, slowing down slightly. This initiates a backward-propagating avalanche of stopped cars that spontaneously forms a traffic jam. Only those cars on the front line are able to take advantage of the now-lower density of cars ahead of them and accelerate out of the jam. Those in the middle and back end of the stalled pack must wait for this leading edge to diffuse back to them before they can resume their journey. Even when the car in front has moved on, the next car does not instantly start accelerating, as its driver also has a finite response time. Of course, while the stopped cars are waiting for the unblocked front edge of the cluster to reach them, new cars are stopping and joining the jam from the rear.

If you were to drive at a steady, uniform speed, keeping pace with the average flow of traffic, then not only would your chances of getting stopped in a spontaneous jam diminish, but you would actually make it easier for the drivers following you. By driving at a steady speed and deliberately *not* shrinking any gaps, you will avoid joining a dense cluster of cars ahead; and if a jam does form, you will give it time to evaporate before you reach it. In essence, you should elect to drop out of the tail end of a traffic wave, smoothing out the density of cars on the road. If all cars drove at the posted speed limit, or at the same average velocity, then the numbers of cars that a highway could carry would climb to the theoretical maximum, and you, along with every other driver, would reach your destination faster.

Your car inches slowly forward; you are completely boxed in on all sides. It must be an accident, you think, and a pretty bad one, to tie up traffic like this. Looking to the left and right, you cannot see any emergency vehicles on the shoulders. Finally, the cars in front of you begin accelerating to highway speed, and you find yourself out of the jam. In another two miles you reach your exit and, continuing to follow the GPS directions, arrive at your doctor's building. There is a long line of cars waiting to enter the pay-parking ramp across the street, which you'd like to avoid. To your surprise, a car pulls out of a curbside parking space right on your doctor's block. It is a tight fit, but your automobile's **self-parking feature** *smoothly maneuvers you into the parking space.*

In order to take human miscalculations out of parallel parking, the self-parking system must know exactly where the car is located relative to other cars and obstacles.* This is not the same as the GPS, for you need a more granular situational awareness in order to avoid hitting other vehicles. An onboard computer communicates with the power-steering system, turning the wheel just so, taking into account the speed with which the car is backing into the space (this speed is still under the control of the driver in many such systems). While this feature has only recently become available, the physical mechanism by which a self-parking car can avoid colliding with other

* While many would resist handing over the driving of their car to an onboard computer navigation system, I for one welcome our robot overlords taking over the task of parallel parking.

objects has existed since World War II, with the development of radar and the proximity fuse.

The word "radar" is an acronym for *radio detection and ranging*. Radar involves the emission of a pulse of electromagnetic waves in the radio portion of the spectrum that is reflected from an object and subsequently detected by an antenna, measuring distances by the echo technique. In the case of a self-parking car, a signal pulse is emitted by an antenna/receiver mounted on the vehicle's bumper. This pulse is reflected from an object and then travels back to the receiver. By noting the time between the emission of the outgoing pulse and the receipt of the reflected pulse, and knowing the speed of the signal, you can easily calculate the distance between the car and the object. Multiple receivers positioned on the car's front and rear bumpers provide accurate determinations of its position relative to other automobiles. However, radio waves have large wavelengths and thus can miss small objects that may not scatter the waves as effectively as large objects do. Some self-parking systems thus employ "lidar" (similar to radar but using shorter-wavelength, visible-light laser beams) or ultrasonics (short wavelength sound waves) in order to detect pedestrians and other obstructions smaller than a car.

The foundation for self-parking (and self-driving) automobiles can be traced back to the proximity fuse. An important application of radar in World War II was embedded in antiaircraft artillery that enabled a shell to explode once it came within a certain distance of its target. Inside the shell was a glass container filled with an electrolyte (a fluid that is electrically conductive). The container would break open due to the violent shaking that occurred during firing; this closed a circuit that would then emit a radar signal. A separate circuit in the shell would detect the reflected waves, whose signal would grow stronger as the shell approached its target. Typically, when the shell was seventy-five feet from its objective, the reflected signal would be strong enough to induce a voltage to be sent to the triggering charge, exploding the bomb. As a result, a perfect bull's-eye wasn't necessary to damage the target—important when trying to hit fast-moving targets. Thanks to the radar-triggered proximity fuse, close could be close enough; and the same physics also tells you when you are close enough to the cars in front of and behind you as you parallel park.

Your car has managed to park you snugly in the space, and you try to hurry to your doctor's office, though the pain in your left ankle keeps you from moving too fast. At the parking kiosk you enter the number assigned to your parking space, slide a credit card into the reader, and automatically pay for the street parking. You accept the charges, and the machine prints out a paper receipt, indicating in bold, large type the time you need to be back. Heading back to your *car, you grab your tablet and lock your car using the **keyless remote control.***

Your remote entry system for your automobile operates as a one-way E-ZPass. When you press the lock or unlock button, your keyless entry fob sends a radio signal to a receiver in your automobile that then operates the power locks in the doors (or even starts the vehicle). The receiver in your car detects the radio signal sent from the fob using the same physics underlying a radio tuner: resonance.

We live in a sea of electromagnetic waves, from sources as far away as the sun (which sends us light of all frequencies, those of greatest intensity being in the narrow frequency slice we call "visible light") to sources as near as the infrared light emitted by our own bodies. We typically never think about all the electromagnetic waves around us, except when we can't get enough bars on our mobile phone to make a call. A radio tuner, or the receiver in your remote entry system, must be able to ignore all of these frequencies, except for the particular one corresponding to the station you have selected. It does this by using a "tuning fork" as a receiver.

A tuning fork consists of a long piece of metal bent into a U-shape. When struck, the two long sides of the U-bar vibrate at one particular frequency, creating pressure waves in the air. The tuning fork thus converts a rapid, one-time mechanical disturbance into sound at the unique frequency of the fork. Changing the length, mass, or shape of the bars in the fork will change the frequency of sound it emits. Another way to get the bars of the tuning fork to vibrate is to expose them to sound waves at the exact natural frequency of the fork. Any other frequency won't cause the tines of the tuning fork to hum. This phenomenon, where you excite an oscillator at its natural frequency, is called "resonance."

Your radio tuner and the remote entry receiver in your automobile (or in your garage door opener or E-ZPass system) use the same principle in order to home in on the one frequency that is relevant, and disregard the background sea of other electromagnetic waves. The electronic circuitry is more complex than a simple pendulum or tuning fork (though in the hierarchy of electronics, the radio tuner circuit is actually rather simple), but it functions essentially the same way. When an electromagnetic wave is detected at the selected natural resonant frequency of the circuitry, a much larger voltage is induced than for any other frequency, and this larger voltage is then sent to the rest of the system, which processes the information encoded in the electromagnetic wave. To change the resonant frequency of a pendulum, you can change the length of the string, while changing the size and shape of a piezoelectric quartz crystal changes its natural frequency. When you change the radio station, an alteration of the receiver's circuitry varies the resonant frequency of the radio tuner.

All automobile remote entry systems in the United States use the same radio wave frequency of 315 Mega-Hertz. Since all systems use the same frequency, security relies on knowing the right randomly selected twelve-digit code that is sent at the same time as the procedural command. Only if the receiver detects the correct code number will it follow through with the other signal sent (by opening the trunk or locking the car doors, for example).

Couldn't someone with a receiver pick up the signal from your remote entry fob (just as two radios can listen to the same station) and thus intercept this twelve-digit code number? They could, but the next time you use your remote entry fob, a *different* twelve-digit number is used as a password. So using the last twelve-digit code won't help you open the car door. The receiver knows to accept this new twelve-digit number because the mathematical formula that your fob uses to generate it is the same one used in the receiver, and they both started off with the same "seed." Thus, only one unique twelve-digit number will be recognized by the receiver. The next time you press the remote entry fob, it generates a *new* twelve-digit number, using the last twelve-digit value as the seed. Intercepting one signal will thus not enable you to open the car a second time.

What if you accidentally press the fob when you are far away from your automobile, so that the receiver in the car does not know that the twelve-digit password has been updated? For just such a case, the receiver also scans the next 256 twelve-digit numbers the algorithm would generate, in order to find the right one.

CHAPTER THREE

You Go to the Doctor

*You enter the high-rise building where your doctor's office is located and head over to the elevator bank. The office is on one of the top floors, and you are concerned about additional delays. But soon, quicker than you anticipated, the elevator arrives. You feel a momentary heaviness as the **high-speed elevator** begins its rapid ascent to your floor.*

An elevator is essentially a pulley, with a cable that loops over a cylindrical drum. One end of the cable is affixed to the elevator car and the other to a counterweight. The size of the counterweight is chosen to reflect the weight of the car with its average occupancy, typically 40 percent of the maximum capacity of the car. By making the counterweight as close as possible to the weight of the car, we minimize the energy needed to raise or lower the elevator car.

Think about a see-saw. It is much easier for each side to rise and fall if children of equal weight sit on either end. If a heavy adult is on one end, and a small child on the other, additional force is needed to raise the adult. In an elevator, one end of the see-saw is the car with people inside, and the other end is the counterweight. When both are nearly the same mass, then the elevation of one (leading to an increase in its potential energy) is exactly balanced by the descent of the other (with an equal decrease in potential energy). If there is no net change in potential energy for the car/counterweight system, then the work done by the motor turning the pulley is reduced. If the car is completely empty, or at its total capacity, then there will be a net change in the potential energy of the car/counterweight system, and this change in energy must come from additional work done by the motor.

The length of the cable connecting the elevator car and the counterweight has to be just a bit over the height of the building. When the elevator is in the lobby, the counterweight is at the top of the building; conversely, the counterweight is at ground level when the elevator is at the highest floor.

But the cable does not just hang over the pulley. There must be sufficient "grab" so that when the motor turns the drum, the cable moves along with it, causing the car to rise or descend. The pulley is essentially a long cylinder, and the cable holds on to it through simple friction. The more contact surface area, the more rubbing, and the greater the frictional force. This is why the cable is wound several times around the cylinder, instead of being draped just once. Grooves cut into the cylinder in which the cable lies (a pulley with grooves is technically called a "sheave") also increase the friction. The center of the cylinder is connected to the shaft of an electric motor, which can rotate either clockwise or counterclockwise and at variable speeds. In this way, the cable connected to the elevator car can be pulled up or let out.

So that's the basic physics behind how an elevator works—but what do you have to do to get it to go fast safely? Speed is mostly dependent on the motor that turns the top pulley. In a motor, a changing current in a coil generates a magnetic field that is alternately repelled from and attracted to the magnetic field of a permanent magnet or an electromagnet. The stronger this magnet, the greater the torque—the twisting force that the motor can produce—and the quicker it can spin the sheave up to a high rotation speed, which translates to a faster rise or fall of the elevator car.

The safe part takes a little more work. The main rule for elevator safety is to employ a redundant system of redundancy.* One thick steel cable is strong enough to hold up a full elevator car—most elevators actually use four to eight cables. If one or even two were to snap, the car wouldn't drop to the bottom of the shaft.

Even if all eight cables were to snap simultaneously, you *still* wouldn't fall. The elevator car is surrounded on two sides by vertical rails that run the height of the shaft. Rollers from the car grip these rails, keeping the car perfectly vertical (even if all of the passengers stood on one side of the car, the elevator wouldn't tip). These rollers have accelerometers that detect if the car is shaking or wobbling, and apply compensating forces to counteract the vibrations (not unlike anti-lock brakes that can pump much faster and more sensitively than a driver can, following input from the vehicle's accelerometers). There are additional elevator brakes gliding along the rails that will grab those rails if an excessive acceleration is detected. These brakes use electromagnets that force the clamps open—when power is lost, the clamps automatically close onto the rails, acting as a sort of dead-man's grip. These rail brakes have been around for years, but high-speed elevators would generate such frictional heating when they tried to arrest the motion of a fast-moving car that the metal in the brakes would melt. Today, brake clamps are coated with an extremely hard ceramic material that can withstand much higher

---

* That's a joke, son!

temperatures than metals. (These ceramics are also employed in brake pads on certain high-performance sports cars for the same reason.) There is also a "governor" in the rotating pulley, with hooks that will engage stationary ratchets, bringing the cylinder to a stop if it were to rotate at too high a speed. This safety feature is very clever, capitalizing on the fact that at higher rotations, the hooks will fly away from the center of the spinning pulley, until they grab the ratchet, stopping it.

An elevator car moving at over 40 miles per hour inside an elevator shaft needs to expend considerable energy pushing the air out of its way. Specially designed, curved light-metal covers at the top and bottom of the car improve its aerodynamics, reducing the energy needed to move through the air. Air pressure is a concern inside the car as well, as a trip from the top of a tall skyscraper to the lobby leads to the same unpleasant pressure on your eardrums as experienced in airplane landings. Small fans inside the elevator car adjust the air pressure as it begins its descent, so that when you arrive at the ground floor, the pressure is already equilibrated.

Another physical phenomenon associated with airplane rides is also experienced in these high-speed elevators. In order to reduce travel time, the motors attached to the sheave must start spinning the drum very quickly. That is, the rotational acceleration is high, translating to a large linear acceleration of the car. The feeling of heaviness or lightness you experience during these accelerations is a perfect illustration of the "equivalence principle" of the general theory of relativity: in a closed room, such as an elevator car, it is impossible to distinguish between the effects of acceleration and gravity. When the car begins its ascent, there is a large acceleration upward, and the elevator floor exerts an additional upward force (in addition to the force balancing your normal weight), so you feel as if your weight had increased. During most of the nonstop ride to the top, the car travels at a uniform speed, so you feel your normal weight. But near the top, the car needs to decelerate and the elevator floor exerts a large downward force, so that your apparent weight is reduced. This is a mini-version of when certain airplanes perform a controlled parabolic arc, at the height of which the passengers experience a brief period of feeling weightless.

*You arrive at the doctor's office, check in, and are handed a clipboard with a series of forms, asking for information that you have provided every time you have visited. Despite the fact that there has been no change in your medical insurance, the receptionist asks for your insurance card, makes a photocopy of both sides, and asks you to be seated. You sit down in the waiting area, turn on your smart tablet, and enter your passcode. The screen opens to the home page, displaying the time, temperature, and a series of app icons. You ask the receptionist if **wi-fi** is available in the waiting room.*

She provides the network password, and your tablet soon shows the concentric arcs indicating that you are connected to the office's system. You finger-swipe to another page full of app icons and select one for your airline. You checked in online last night, but now you want to make sure your flight is on time. The screen fills with the airline's website. You tap FLIGHT STATUS, *and the site requests the flight number or the origin and destination. You can recall your flight number without looking it up, as it is the same as the year you were born. You enter this information, there is a slight pause, and then the screen informs you that your flight is* ON TIME, *as well as providing your gate information. You're quite familiar with this airport, and recognize that your gate is a fair distance from the TSA station. You are hopeful that you won't have to run to your plane (not on this ankle!). Just as you are about to ask about the wait, the nurse appears and calls your name.*

The simple act of visiting the airline's website by tapping the appropriate icon involves a wide variety of physical phenomena. First, whenever your tablet accesses a website, it is in essence making a phone call. Instead of hearing a dial tone that indicates the phone line is active, you look on a tablet (or smartphone or laptop computer) for a series of concentric arcs that confirm the device has detected the particular radio-frequency signal of the nearest wireless router. (The physics is the same when using your data plan, so we'll stick with wi-fi.) The website address you type into the Internet browser is actually a nickname for a numerical address that is roughly the equivalent of a phone number. For a telephone call, an exchange determines the best way to route the signal from your phone to the one you are calling. For websites, there are certain designated servers that have the location of the host computer sponsoring the requested site. The host computer stores all the information (formatting and content) that is available to you when you visit the site.

Your smart tablet communicates via radio waves (wi-fi)* with a computer that sends the requested sequence of voltage pulses containing the information for the website you wish to access. The transmission of information in a landline phone call used to be done in an analog format. Now most phones, as well as your smart tablet, handle vast streams of digital information, that is, a sequence of ones and zeros. Which is fortunate because, while the semiconductor processors in a tablet or smartphone are very fast, they themselves are not very smart.

The difference between analog and digital is the difference between something that can change continuously and something that can be altered only in discrete

---

* While it sounds similar to "hi-fi," which is shorthand for a high-fidelity stereo system, the word "wi-fi" is not a nickname or acronym for anything.

steps. For example, the sound waves generated by external speakers are analog waves, as there is a continuous variation in their amplitude and frequency. The voltage that told the speaker membrane how to vibrate used to be analog but is often now a digital voltage created by adding many tiny voltages together. The smaller the voltage steps used in a digital voltage, the more closely the digital signal will represent the true, smoothly varying signal when conveted to an analog voltage that should be sent to the speakers.

Even though digital signals are an approximation of the true voltage in an analog signal, they have clear advantages in devices like smart tablets. In the days before digital phones, long-distance communication was difficult and the reception was often poor. This is because the analog voltage sent along the metallic telephone wire was susceptible to noise in the wire (arising, for example, from the random motion of electrons in the wire, which in turn induced voltage fluctuations that competed with the signal you wished to send). A more reliable way to transmit information is to use a digital code, such as the dots and dashes representing numbers and letters in a Morse-code telegraph system. This way, if certain digital bits are lost, the message's content often will still be understood by the receiver. Plus, if you don't want anyone to intercept and read your transmitted message, it is easier to encode a sequence of ones and zeros than a full analog signal, which is great for online commerce.

Your screen is made up of a sequence of closely spaced dots called "pixels," and the digital code received by the tablet tells each pixel whether to be bright or dark, and which color filters to activate. Each pixel is roughly ten times bigger than a single cell in your body, so we do not see the image's intrinsic graininess. The net result of all of these bright and dark dots is a high-resolution image. Most tablets have approximately 1,500 pixels along one length of the screen, and 800 to 1,000 pixels down the other side. All these dots add up to over 1 million pixels, or a "Megapixel."

Your tablet also monitors the status of each of these million pixels. By touching the screen in a specific spot, such as a box labeled FLIGHT STATUS, you send a signal to the processor in the tablet. You are initiating a conversation with the computer that is sending the website's digital data stream to your tablet. The computer hosting the website responds by sending a new set of instructions, and the display on your tablet changes, asking for your flight number. When you enter this information, the host computer sends another data stream, and the display changes once again, telling you that your flight is scheduled to depart on time.

The nurse takes you back into a suite of rooms, weighing you and measuring your height in the hallway before tak-

*ing you to an examination room. You sit in a chair as he goes through a standard checklist of questions about your medical history, entering your responses on a small laptop he brought with him. He then takes your blood pressure, using a device that hasn't really changed in decades. Instead of monitoring your pulse with a stethoscope, he places a plastic clip, like a small clothespin, on one of your fingers. Having recorded your blood pressure and pulse, the nurse next takes your temperature, using an **electronic thermometer**. Rather than keeping a thin tube of glass propped awkwardly under your tongue for several minutes, you hold the electronic thermometer in your mouth for only a few moments before it indicates that your body temperature is normal. The nurse notes this on your electronic chart and, as he leaves, informs you that a technician will be by in a minute to take you for an x-ray.*

The old-school glass thermometers consisted of a small hollow cylinder, sealed at both ends and partially filled with a fluid (either mercury or red-dyed alcohol). One end of the thermometer was brought into contact with an object (such as your body, at a point underneath your tongue), and if its temperature was greater than room temperature, the liquid would expand, rising in the cylinder.

How do we know that the height of the liquid in the

glass cylinder corresponds to a particular temperature? If you put the end of the thermometer in an ice bath (0°C, or 32°F) and note the height of the liquid, then put it in a pot of boiling water (100°C, or 212°F), you'll notice the liquid has risen—and should do so smoothly and uniformly as the temperature increases. Mercury was best suited for this, as it stays a liquid down to −38°F and doesn't boil until 674°F. Red-dyed alcohol boils at 173°F —good enough when monitoring a person's temperature.

But why does the height of the liquid, whether it is mercury or alcohol, change with temperature? Even if objects expand when heated, won't the glass cylinder expand as well, so that there should be no net change in the height of the liquid? Temperature represents the average energy-per-atom in an object, whether solid, liquid, or gas. For most materials there is a small extra component to the force between atoms (called an "anharmonic term") that provides a bit more pull in one direction over the other. This leads to a small but real net relative displacement of the atom with rising temperature. The amount that a material's volume increases when warmed depends on the detailed arrangement of the atoms and the precise nature of the forces between them that hold the material together. Glass has a much smaller anharmonic term than either mercury or alcohol, so as you warm up the liquid in a thermometer, that liquid expands faster than its glass container.* The conduction of heat through glass is poor (which is why it makes a decent thermal insulator), so a thermometer consisting of a fluid-filled glass tube has to be left in thermal contact under the tongue for some time to make sure the mercury or alcohol has reached the same temperature as the patient.

Any system can be used as a thermometer as long as it has an easily measured physical property that changes uniformly with temperature. This includes not just the structural properties of solids, but their electronic properties as well. Metals are not very good electrical conductors at higher temperatures, when their atomic vibrations increase. The shaking atoms can scatter the electrical current, just as imperfections increased the resistance of the toaster wire. Alternatively, semiconductors are not very good conductors, as all of their electrons are locked into chemical bonds between the

---

* Engineers must pay very careful attention to the thermal expansion of materials, so that components of a device such as interlocking gears do not change their separation with temperature. But thermal expansion can sometimes be used to advantage. For example, think of a long metallic strip, where one side is made of one metal and the other is composed of another metal. When turned into a coil, this bimetal strip—due to each metal having a different rate of thermal expansion—will unwind or tighten itself as the temperature changes. Older models of thermostats used the winding and unwinding of such coils as a function of temperature to mechanically move a switch, increasing or decreasing the heat sent from the furnace, or to turn a toaster off when a preset temperature is reached.

atoms. Raising the temperature allows some of the electrons to momentarily leave their chemical bond confinement, dramatically lowering the electrical resistance. By calibrating how their electrical conduction varies with temperature, one can use measurements of a metal's or semiconductor's resistance as a thermal sensor.

Another common technique to measure temperature electrically involves a device called a "thermocouple." When two dissimilar metals are fused at a single point, a voltage can develop across the connection that is sensitive to temperature. Different metals will have different numbers of free electrons per volume, so one metal can have more mobile electrons than the other. When the metals are connected, some electrons from one metal will move over to the other, just as gas atoms will expand from a high-pressure region into a lower-pressure volume. Before they were connected, each metal had the same number of negative electrons and positively charged ions, but after they are fused there is a net electric-charge difference across the junction. A small but measurable voltage thus forms between the two metals. As the temperature of the metallic junction is changed, some electrons will move from one side to the other, so this voltage will vary as the temperature changes. Metals are also very good thermal conductors, so this electronic thermocouple-based thermometer provides an accurate temperature reading very quickly. The voltage generated by the thermocouple junction, or from the temperature-dependent variation in the resistance of a metal or semiconductor, is detected by a sensor chip in the electronic thermometer, which instructs a liquid crystal display (LCD) to show the measured temperature.

The technician takes you to another room, with a slightly wider examination table, the standard sheet of white paper spread across it. She asks you to take off the shoe and sock of the foot that has been bothering you. The technician slides a black rectangular platter onto the table and asks you to place your ankle on top of it. You have to spread out at a funny angle to orient your ankle in the proper position. The source of the x-rays is lowered from the ceiling of the room. The technician makes sure your foot is pointing the right way, then places a lead-lined apron over your torso and midsection. She briefly steps behind a protective wall and turns on the x-ray machine. There is a slight buzz, and the **x-ray scan** *is completed. The technician can tell right away that your foot moved while the scan was being taken, so she comes back in to readjust your leg. You make an effort to remain completely still, and this time the scan is successful. You ask how long it will take to get the image developed, and the technician shows you on a computer screen the* *high-resolution x-ray of your ankle. She informs you that the image has already been sent electronically to your doctor, and after you put your sock and shoe back on, she leads you to your examination room. She tells you that the doctor will be in to see you shortly.*

X-rays are electromagnetic radiation, just like radio waves, microwaves, visible light, and ultraviolet light. The most common method of generating them employs the physics principle that we've invoked several times before—changing electric currents create changing magnetic fields. In an x-ray tube, a wire ejects electrons at one end of the tube, and a positive voltage on a screen mesh at the other end attracts these electrons and accelerates them up to a very high kinetic energy. The whole apparatus is enclosed in a glass jar from which the air has been removed (this minimizes scattering of the electrons by the air molecules), a forerunner of the vacuum tube. Most of the electrons do not strike the wires in the positively charged mesh, but instead barrel through the screen's openings and then reach another plate, with a negative voltage, that deliberately and rapidly slows them down. Any change in the electrons' motion results in the emission of electromagnetic waves (think of the red glow coming from the toaster wire). Under the right circumstances, the rapid deceleration of the speeding electrons yields light with a very short wavelength of about the size of a single atom, termed x-rays.*

As x-rays move into a solid like your ankle, they interact with the electrons in the atoms in the material. These electrons scatter the x-rays like randomly oriented mirrors, so while some of the x-rays continue on in a straight line to the detector plate, most are reflected in other directions. The more electrons in an atom, the more effective it is at scattering the x-rays. Carbon has only six electrons, so it allows a lot of the x-rays to pass through it. Lead has eighty-two electrons, and is more effective at scattering x-rays. This is why x-rays do not penetrate lead. Gold has seventy-nine electrons, and is just as opaque to x-rays as lead. (For obvious reasons,** the protective aprons used in a doctor's office to block unwanted x-ray exposure are not made of gold.)

A water molecule has only ten electrons (eight from the oxygen atom, and one each from the two hydrogen atoms) and therefore does not significantly scatter x-rays. Since we are mostly water, neither do we—and x-rays easily pass through our bodies. A calcium atom has twenty electrons and is much more effective at scattering x-rays as water molecules. Our skeletons show up bright on x-ray images, as the atoms in our bones reflect the x-rays

---

* Physicists refer to this light as *Bremsstrahlung,* which is German for "braking radiation."

** Gold shielding would be insanely expensive, and doctors using it would likely be continually robbed!

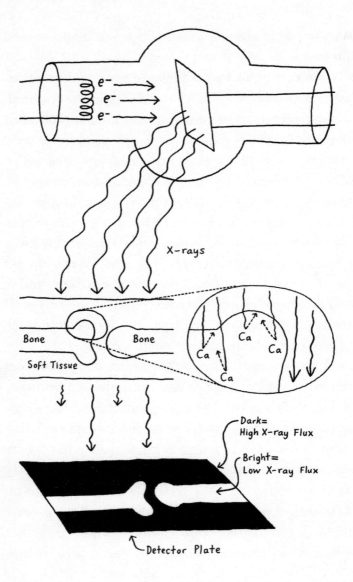

FIGURE 3

away from the detector plate. Any region that is missing atoms, such as a cavity in your tooth or a fracture in a bone, will not scatter the x-rays, and the resulting higher x-ray flux will show up as a darker region on the detector plate. Even in soft tissue, a section with a higher density of atoms will have more reflectors than the surrounding material, and this mass can be detected in the x-ray scan. In order to improve contrast, some x-ray scans require the ingestion of a fluid containing compounds of iodine (which has 53 electrons) or barium (barium alone has 56 electrons, and barium sulfate, used in scans of the digestive system, has 104 electrons). The incorporation of these molecules improves their ability to scatter x-rays and enables the doctor to obtain an accurate image without having to increase the x-ray beam intensity to dangerous levels.

While being able to view structures beneath the surface is a great boon to doctors and diagnosticians, there is a limit to the amount of information that can be gleaned from a single two-dimensional snapshot. A full, three-dimensional image is ideal. A given mass inside your body that is scattering x-rays may be large in two dimensions but skinny in the third, like a pancake. If this "pancake" were oriented so that the x-ray image showed it edge-on, it would appear to be a long but narrow wedge and might not seem a cause for concern. A full 3D image would give an accurate representation of the scattering structure's true size and extent. Of course, one cannot just rely on two orientations at right angles to each other (like a mug shot), as the mass may not cooperate by lying exclusively in one plane or the other. To construct a full three-dimensional image using x-rays, we need to take multiple images, with each scan acquired at a slightly different position. The process of combining a series of x-ray images in slices to create a three-dimensional image is known as "tomography." When it's performed with a computer to resolve the shadows and extract a complete three-dimensional image, this type of x-ray imaging is called a CAT scan (for "computer-aided tomography") or a CT scan.

"Tomography" (from the Greek for *to write by sections*) is the process whereby one takes successive images using a probe that can penetrate beneath the surface (such as x-rays or sound waves). As the slices systematically move through the object, a full three-dimensional image can be reconstructed. Think about imaging an orange that sits on a tabletop, using slices that are parallel to the table surface. The first slice at the top would yield a small dot. As the slicing plane moves down toward the table's surface, the images would be of circles that grow in size. Once past the midpoint of the orange, these circles would become smaller, and at the bottom of the orange we would again have a final slice with a small dot. Comparing and combining these individual slices, each one of a circle with varying diameter, you would conclude that the orange is shaped like a sphere. Alternatively, the same process performed on a can of soup would yield a continuous series of imaged circles of constant size, indicating that the soup is stored in a cylinder. These simple cases can be analyzed by hand— similar procedures on a person require a computer to reconstruct the image slices. The greater the number of separate slices, the more accurate the resulting image will be, but this also involves longer exposures to x-rays than in a single, two-dimensional scan.

The doctor arrives a minute later. She greets you and asks how you are doing. Fine, you reply, except of course for the ache in your left ankle, the reason for this morning's visit. Pulling up an image from your file on her laptop, she shows you the x-ray of your ankle, pointing out the small amount of arthritis you have. When you protest that you are too young for arthritis, she postulates that this is the result of an old athletic injury. Perhaps you suffered a hairline fracture years ago in school and thought it was only a swollen ankle. She is optimistic that an injection of cortisone will reduce the pain. The shot will be administered at the precise location of the current inflammation, using **ultrasound imaging** *to guide the needle. The procedure takes about an hour, and the ultrasound machine is fully booked for the day. You'll just have to carry on as you have been, for the remainder of your day.*

The doctor tells you that you can make an appointment for the cortisone shot with the receptionist, and she gives you some extra-strength analgesics to take in the meantime.

S ound is a wave phenomenon, and, perhaps surprisingly, it can also be used as a visual tool. In medicine, sound waves can be used to provide visual information about a patient's interior. While electromagnetic waves can propagate through the vacuum of empty space, sound waves cannot exist unsupported by a medium—either a gas, a liquid, or a solid. In ultrasound imaging, also known as "sonography," that medium is your body. The fact that the speed of sound varies with the density of the medium is the basis of ultrasound imaging.

Sound waves are periodic variations in the density of a medium. When you strum a guitar string, you cause it to vibrate at a given frequency that is determined by the string's properties—tension, mass, and length. Think of the string vibrating, with the center of the string moving up and down at one particular frequency. As the string moves up, it collides with air molecules and pushes them in the direction the string is moving. The air molecules pushed by the string pile up, leaving a region that is relatively depleted of air molecules behind it. When the string swings downward, it disturbs the air in the direction of its motion. On the next upward swing, the string again causes the air molecules to pile up into a higher-density region. The vibrating string thus induces a periodic modulation in the density of the air above and below it that propagates outward from the string.

Sound also travels through liquids, such as water, via the same physical process, and thus sound waves can propagate through us. The speed of the waves depends on the properties of the fluid and, all other factors being equal, the denser the medium (that is, the closer together the atoms are jammed next to each other) and the weaker the connections between the medium's atoms, the slower the sound waves will travel through it. When sound waves strike the interface between two media in which the speed of sound changes, some of the waves will be reflected at this boundary. By measuring the changes in reflection as the position of the source of the sound waves is varied, the ultrasound device can map out the density variations inside a person.

Typical ultrasound generators used in medical imaging applications have a frequency of roughly two to three million cycles per second, well above the detection range of even a person (or dog) with excellent hearing. Sound waves with this high a frequency have a correspondingly short wavelength, which is desirable for imaging. The shorter the wavelength, the smaller the features that can scatter the waves (just as in the case of lidar used by the self-parking car). If the separation between adjacent peaks in the wave is three feet, then one will not detect significant changes in the wave if it strikes an object that has a length of an eighth of an inch.* To detect small-scale structures in the body, ultrasonic imaging devices use sound waves with a wavelength close to that of visible light. But unlike visible light, these sound waves can pass through the surface of a person, to be reflected from the internal organs.

In an ultrasound imaging system a piezoelectric material generates the sound waves. By using an alternating voltage that smoothly oscillates between a positive and a negative polarity, we can cause a piezoelectric material to alternate between compressed and expanded volumes, generating sound waves in the same manner as does a vibrating guitar string. The piezoelectric material, vibrating at very high frequencies, generates pressure waves in any medium in contact with the crystal—such as the fluid in your body. The jelly placed on your skin in preparation for an ultrasound scan is there to improve the transfer of sound waves from the piezoelectric transducer to your body (called "impedance matching"). The waves move into your interior, being reflected at any in-

---

* This is why one cannot view individual atoms or smaller molecules using an optical microscope. The wavelength of visible light is roughly one thousand times larger than the diameter of an atom. In order to image individual atoms, one must employ quantum mechanical tricks, as in an electron microscope.

terface where there is a change in density that results in sound-wave speed variations. The reflected waves are then detected by a piezoelectric material, which is subject to the oscillating pressure waves compressing and expanding the solid, in turn generating a voltage that is processed and transformed into an image. These interfaces are somewhat subtle, which reduces the contrast in the resulting ultrasound image. Varying the position of the source of the ultrasound waves yields reflections from a variety of angles, performing a simple tomographic imaging procedure. The engineering and data processing that go into obtaining an image are very complex, but at its heart the physics of these devices is essentially the same as what's at work when, looking out into the night through a window, you see your reflection in the glass.

*You stop by the receptionist's desk to make an appointment for the ultrasound and the cortisone injection. However, when the receptionist checks the doctor's orders on her computer screen, she misreads the notation and thinks that you need to schedule a **magnetic resonance imaging (MRI)** scan. You realize that there has been some miscommunication when she asks you at which hospital you would like to schedule the procedure, and if you have any tattoos. After a minute or so, it's sorted out that you do not need an MRI. Checking your* *calendar on your smartphone, you agree to next Monday for your cortisone injection.*

M agnetic resonance imaging uses the fact that the building blocks of atoms—the electrons, protons, and neutrons—each have a small, built-in magnetic field, like a little bar magnet with a north pole and a south pole.* Different elements have different numbers of protons and neutrons in their nucleus, and thus have different nuclear magnetic fields. How can you measure the strength of the magnetic field of each magnetically active atom in your body? By getting all of the nuclear magnets to point in one direction and seeing how much energy it takes to flip them 180 degrees. When you lie down inside a very large electromagnet, with a big magnetic field whose north pole is pointing toward the ceiling (for the sake of argument), then the nuclear magnets in your body will want to align with this external field, just as a compass needle will be drawn toward a magnet brought near it.**

---

* In fact, originally MRI was called NMRI, for "nuclear magnetic resonance imaging," but for marketing reasons the "N" was dropped from the name.

** Purists will note that, technically, the nuclear spins do not fully align with the external magnetic field, but precess around it, like magnetic Gene Kellys in *Singin' in the Rain*, spinning at an angle around the external magnetic lamppost. Purer purists will further note that, technically, this is a quantum mechanical process, and the sensitivity of the energy of a nuclear magnet to an external magnetic field is known as the Zeeman effect. But if you know all this—why are you reading this book?

To be precise, the nuclear magnetic fields have their lowest energy if they point in the same direction as the outside magnet, and they'll have their highest energy if they are aligned in the exact opposite of the magnet's direction. This is the reason you can't take any metal or credit cards into the room where an MRI is active, for the massive electromagnet of the MRI device can exert a large force on these objects. Similarly, the changing magnetic field of the MRI can cause heating in any metal inside your body, or even on your skin if there are metallic inks used in your tattoo.

Now, the magnetic field of the nuclei in your atoms is extremely weak; even when you place them in a very strong external field, it doesn't take much energy to rotate them so that they are pointing away from that field. This is great news, as the light needed to flip the nuclear magnets is at the low end of the electromagnetic spectrum—radio waves that easily pass through your skin and inside your body.

When treated quantum mechanically, light consists of discrete packets of energy termed "photons." The energy of any form of light is determined by its frequency. (This was first recognized by Albert Einstein back in 1905, the same year he developed the special theory of relativity.) A feature of quantum mechanics is that nuclei and atoms can absorb a photon only if it exactly corresponds to the separation of their energy levels. If the photon has too little or too much energy, absorption can't happen. The nuclei of different elements will have different magnetic-field strengths. The radio wave photon energy needed to flip a nuclear magnet in an external magnetic field will therefore differ for each element, and the frequency (and hence energy) of the absorbed radio waves indicates which atoms are present.

Radio wave frequencies can be tuned very precisely, as anyone knows who has tried to listen to a radio station broadcasting at 100.0 MegaHertz when the radio dial is set to 100.3 MegaHertz. The dominant MRI signal comes from water molecules that fill every cell in your body. Pairs of protons (or neutrons) join together, with the north pole of one facing the south pole of the other, so that for each pair there is no net magnetic field. The oxygen atom in water has an even number of protons and neutrons, and consequently its nucleus is not magnetic. The hydrogen atoms in H_2O are essentially isolated protons, and their internal nuclear magnetic field will interact with the MRI device's magnetic field. These hydrogen atoms are the main source of the detected signal.*

In order to obtain contrast in an MRI scan, the magnetic field is deliberately made to be very weak on one side of your body (say, the left side), increasing uniformly in strength across your torso, and strongest (in this ex-

---

* There are other atoms with net nuclear magnetic fields in the body, such as calcium, sodium, phosphorus, nitrogen, and isotopes of carbon and oxygen (which have more neutrons than protons—more on this later).

ample) on the right side. So the nuclear magnets require very little energy to flip their orientation on the left side of your body and more energy to flip on the right. By noticing how many of the incoming radio waves are absorbed as their frequency—and hence, their energy—is changed, you can figure out the number and kinds of atoms that are present in the body. Pulling this information together (with the aid of a lot of computing power) creates an image of your interior.

As mentioned, there is a very small difference in energy between when a nuclear magnet is aligned with the magnetic field or opposed to it. Rather than fight against this small signal, MRI devices harness it for image contrast. The system is first bombarded with a pulse of radio-energy photons, until the numbers of nuclear magnets pointing along with the external field and those pointing opposed to it are the same. Now the radio-energy photon stream is turned off, and the nuclear magnets opposed to the magnetic field reorient themselves to the lower energy state (aligned with the field), emitting a radio wave photon.

The time it takes for the nuclear magnet to relax to its original configuration, with slightly more magnets aligned with the external field than opposed to it, is very sensitive to the other atoms near the nuclear magnet. By measuring the time it takes for the nuclear magnets to return to their original state, we can get information about what other kinds of atoms are nearby, and from this we can construct a high-contrast image of organic matter within the person.

Very specific details of the biological state of the tissue, such as whether a tissue mass is benign or cancerous, can be determined by MRI. One recent study exploited the fact that cancerous tumors grow much faster than benign masses, and thus will have a higher metabolism. After being injected with a form of glucose that has an abundance of magnetically active nuclei, the patients' cancerous tumors take up this glucose faster than the surrounding noncancerous matter, and thus on an MRI these malignant regions appear brighter. Nuclear physics not only gives us an image of the body's interior, but also allows us to make crucial, life-saving diagnoses.

CHAPTER FOUR

You Go to the Airport

*You make good time to the airport. You join the line of cars entering the airport, slowing down from highway speed to a more sedate, posted limit of 30 miles per hour. You move over to the lane to park in the long-term general parking lot. You pull up to an e-ticket reader, and an automated voice asks if you would like to save $2 per hour on your parking. The message's goal is to have you pay by **credit card,** rather than requesting a paper ticket. When you leave the parking ramp tomorrow, the exit booths will be fully automated, and you'll reinsert your credit card to pay for your overnight parking. You take out the credit card you use most consistently for travel and push it into the e-ticket reader. You pull the card* *out rapidly, and the wooden barricade, painted with white and orange stripes, lifts up, granting you access to the ramp.*

Information about your charge account is stored on your credit card in different forms: the account number itself is printed on the front of the card; there is a quarter-inch-wide magnetic stripe running along the length of the card; and, in some versions, a memory chip is embedded within the card. Each storage mechanism uses different physical principles.

Raised numbers on the front of the card are the simplest means of storing information, perhaps deceptively simple. While no longer used on some newer credit cards, originally these numerals were stamped into the back of the card, so that the numbers were raised above the card's surface on its front. The utility of having an account number that protrudes from the surface is evident when trying to provide permanent hard copies of a transaction for both the seller and the buyer. In this case some sophisticated materials science was utilized, based on the rather antique technology of carbon paper.[*] Today, carbonless copies employ special paper containing tiny hollow, spherical particles that are filled with

---

[*] Originally, carbon paper was a waxy sheet of paper embedded with carbon-black particles and sandwiched between two plain sheets of paper.

ink. The shell of these particles is not very strong and can be easily broken when one presses firmly on the paper. In the case of the credit card imprinter, the paper is placed atop the embossed account numbers on the card, and then a cylinder is rolled over the paper and card. The height of the roller is set so that it will press on the paper only when it passes over the protruding account number, breaking open the ink-filled particles and permanently recording that your particular credit card was used at a given store.

Your account number, along with additional information, is also stored in the magnetic stripe on your credit card. The stripe contains a series of magnetic regions, whose north poles point in either one direction or the opposite. Magnets are ideal for recording information, as long as one needs to keep track of only two numbers. The way to represent any number using only two inputs (such as north pole or south pole up, or the current in a transistor being high or low) involves "binary numbers." Since binary information is the foundation of all modern digital data manipulation, let's take a moment to explain how it works.

Our counting system, based on the number 10, is reasonable and relatively straightforward, but there are many other ways to represent numbers. If one deals with electrical currents, as in a computer, then the simplest way to manipulate the currents is through a switch, which can be either open (in which case the circuit is broken) or closed (whereby the current can flow). Similarly, small clusters of magnetic atoms in a credit card stripe or a magnetic hard disc can have two orientations, with the north pole facing in one direction or the other. This can be readily mapped to a dot or dash, an "off" or "on," a 0 or 1. The two-digit scheme by which numbers are represented, called "binary," is central to the logic and workings of computers.

Think about making change when you have a limited number of currency denominations. Imagine we have 1's, 10's, 100's, 1,000's, and so on, but no other bills. That is, we start with $1, and each higher denomination is ten times greater than the preceding value (this is called the "base-10 system"). To generate $42, we need four 10's and two 1's. If, in another example, I had two 100's, zero 10's, and four 1's, then I'd have $204. We do this so often that we don't pause to think that we are using a base-10 counting system, and that writing "204" is a way to represent the number of 100's, the number of 10's, and the number of 1's. The drawback to a base-10 system is that it does not work if I have only one bill for each denomination. This is the situation when using magnets that have only two orientations (north pole up or north pole down) to represent numbers. If I were limited to only one single, one ten-spot, and one hundred-dollar bill, then I could represent 101, but not 204 or 42.

Now imagine a different form of currency, where I

have 1's, 2's, 4's, 8's, 16's, and so on. Once again, we start with 1, but now each higher denomination is generated by multiplying the preceding value by 2. This system of counting is called "base-2," or binary. I can still generate any dollar amount I want—for example, 21 is 16 + 4 + 1. Just as in the base-10 system, I start with the ones on the far right, and add denominations to the left. So I would write "21" as 10101; that is, from right to left, one 1, zero 2's, one 4, zero 8's, and one 16. While this is longer and more awkward than the base-10 system, it has one big advantage: even if I had only one bill for each denomination, *any* number can be represented.* Similarly, a binary code can be assigned for the letters of the alphabet, and other abstract symbols.

For the magnetic stripe on your credit card, you can read this information using Faraday's law, which states that changing magnetic fields generate electric currents. In most credit card readers, you mechanically pull or swipe the card through the scanner, causing the magnetic regions to pass by a coil (or some other magnetic sensor).** If the magnetic region has a north pole on one

---

* If you understand that the number 2 is written in binary as "10," with 0 in the ones column and 1 in the twos column, you will be able to appreciate the following "nerd humor": *There are 10 kinds of people in the world—those who understand binary and those who don't.*

** Many stripe readers detect the magnetic regions using a "read head," similar to that used in your computer's magnetic hard drive, where the current through the detector is sensitive to the nearby magnetic field. I've used a coil here as a simpler example.

side and a south pole on the other, then the induced current will flow one way and then reverse direction as the region passes the coil, and be read as a "1." It's the changes in the magnetic field threading the coil that induce a current, and if the magnet has only a north pole or south pole facing the coil, the current will circulate in just one direction and be read as "0." While the actual configuration of the credit card reader is a bit more involved than a simple coil, and the single black stripe on the card actually consists of three distinct bands, each of which contains its own set of information, the underlying physics is the same as described here.*

The third way the account information is incorporated into your credit card is via a small chip that communicates by radio waves with a receiver in the credit card reader. This chip, often termed an RFID, uses radio frequency (RF) light to provide identification (ID). The chip in your credit card communicates with the card reader using the same physics as the E-ZPass system, with more advanced security protocols than your garage door opener or keyless remote entry system. Most of the chips in credit cards use a system called "near field com-

---

* This mechanism is the same as that by which information encoded on magnetic tape—such as in a music cassette or a reel-to-reel tape player—is converted to electrical voltages that, when amplified and processed, cause audio speakers to vibrate. A more sophisticated magnetic storage of information is employed in a videocassette recorder (VCR), but the physics is the same.

munication," in which the radio signal dies away at distances of more than four inches from the card (making it difficult for anyone to eavesdrop on your credit card and steal its information), which is why you must either insert your card into the reader or hold it very close to it. Once the terminal at the parking lot entrance has read and interpreted the information on your credit card, it stores this data (in case you try to avoid paying your parking fee) and opens the gate.

*After parking your car, you retrieve your luggage from the trunk and head over to the main terminal. You realize to your chagrin that you forgot to take any cash with you when you left this morning. You would prefer to have some money with you, just in case an emergency arises. You scan the terminal, and find an ATM off to the side of a newsstand. More good news—the ATM is operated by your bank, so you will not be charged a fee for using it. You insert your debit card (taking care that the magnetic stripe is facing in the proper direction), and the screen transitions from giving you the standard welcome to asking for your PIN. You enter the four-digit code. Even though it has been a while since you used it, you correctly recall your PIN, and the **touch screen** now changes again, presenting you with a series of options. You tap the box labeled WITHDRAW MONEY FROM CHECKING, and the screen changes to a series of boxes, listing options for the total amount you*

*wish to withdraw. You tap the box for $100, and the ATM
asks if you want a paper receipt or if it may e-mail one to
you. You select the e-mail option, and in a few moments your
debit card ejects from the machine, along with five crisp $20
bills. You pocket the money and head over to TSA.*

When you touch the screen at the ATM kiosk, your finger becomes part of an electrical circuit. On some screens, the pressure of the touch alters current flow, while other screens sense your own electrical conductivity. The first kind of touch screen uses "resistive" sensing, as the screen actually resists the force of your finger. The area of the touch screen is divided into an ordered grid, like a checkerboard filled with many small squares. The front of the screen that you touch is a transparent, electrically conducting plastic, which is separated by a thin insulator from a type of glass that has been chemically modified so that it is also a good conductor of electricity. When you press on the screen you force the top conductor to come into contact with the bottom conducting glass, closing a circuit and allowing current to flow between the top and bottom layers. An electronic semiconductor chip records the location on the screen (that is, the location of the squares of the "checkerboard") that now carries a current, and processes this information. The benefit of this mechanism is that it works even if you are wearing gloves or using a stylus pen—touch screens that ask for your signature to verify a credit card purchase are typically "resistive." The downside, as evidenced by the moronic scrawl that your signature becomes on these screens, is that their resolution is not very high. Moreover, as you are physically pressing down on the screen, the touch sensors are susceptible to scratches and other types of damage.

When a lighter touch is needed, another type of touch screen relies on your ability to conduct electricity. Underneath your fingertip in this second kind of touch screen is a device called a "capacitor," which is used to store electrical charges. A capacitor typically consists of two metal plates stacked one atop the other, like the slices of bread in a sandwich. Negatively charged electrons are spread over one of the metal plates, and the other plate develops an equal positive charge. The space between the plates is filled with an insulator (the filling in this "sandwich"). This arrangement is similar to the two conducting sheets separated by a thin insulator in the resistive touch screens—only here we don't have to physically smush the electronic sandwich to register where you are touching the screen.

As before, the touch screen is divided into squares. Each small square on the screen is a capacitor, connected to other circuitry beneath it in the device's interior. When the charge stored on one of the plates in a square is changed, its circuitry registers this modification, and information is then sent to the computer chip waiting for an input. Humans are actually pretty good carriers of electrical charge; think of how you pick up excess charges rubbing against a nonconducting surface such as a carpet. When you touch a particular location, you add your personal capacitance to the circuit; this change in the net capacitance is detected, and an appropriate signal is sent to the processor, indicating which location was selected. In some versions of these touch screens the top conductor is missing, and your fingertip becomes the top plate that completes the capacitor (and modifies the circuit waiting to determine which squares have been selected).* Water (including sweat)** is also conductive, and if it lands on the capacitive touch screen on squares other than the ones you intend to select, it can confuse the sensors.

A basic capacitor consists of two metal plates, but if this were the kind used in touch screens, then you would never be able to see where you should touch. The image on the ATM's screen is projected from a screen underneath the top layer of capacitors that record the region where you touch the surface. The capacitors that make up the touch screen circuit must therefore allow light to pass through them. They employ a special alloy, indium tin oxide, that is both a good conductor and nearly as transparent as window glass. Conducting transparent

---

* One can use a stylus pen on a capacitive-coupled touch screen, but it must be a special pen that has its own capacitor in its tip.

** Pure, distilled water is actually a pretty good insulator. Ions in tap water, rainwater, and sweat can greatly increase its electrical conductivity.

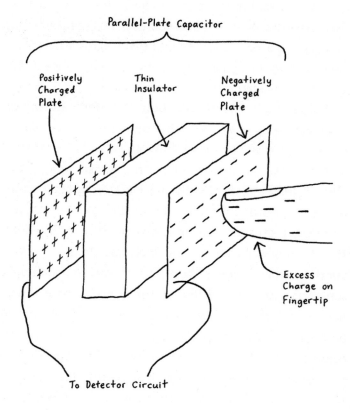

Parallel-Plate Capacitor

Positively Charged Plate

Thin Insulator

Negatively Charged Plate

Excess Charge on Fingertip

To Detector Circuit

FIGURE 4

materials, also used in resistive touch screens, combine two contradictory features—a large density of electrical charges that are free to move under an applied voltage (which makes them good conductors), and a large separation between allowed energy levels (which makes them transparent to visible light that has insufficient energy to promote a charge from lower energy states to the upper energy states).

Some touch screens use light to determine the region you touch or swipe. When moving through something like water or glass, light travels slower than it does in a vacuum or in air. In the denser media, the electric fields of the oscillating electromagnetic waves interfere with the electric fields of the atom's electrons, and this interaction acts as a form of drag on the light. If light moves faster in one medium than it does in another, what happens when it crosses the boundary separating the two different materials? Anyone who has noticed that a straw in a glass of water appears to be broken at the air/water interface has witnessed the consequence of this discontinuity in light speed. The bending of the light beam's path at the interface is a consequence of the light taking the fastest path between two points, and is called "refraction." The property of a given material—whether gas, liquid, or solid—that governs how fast light moves in it, and hence how sharp a bend it makes at the interface between two substances, is called "the index of refraction."

For light moving from a larger index of refraction (where its speed is slower) to a medium with a lower value (where it moves faster), if the angle it makes at the boundary is large enough, the light may bend so much that it doesn't escape the larger-index material. The value of this angle depends on the difference in the indexes of refraction between the two media, and for any larger angle the light will be reflected at the interface, as if it were bouncing from a mirror. This phenomenon is called "total internal reflection," and it is the basis of fiber-optic cable; it's also exploited in ultrasound imaging.

The physical process of total internal reflection is employed in some newer versions of touch screens. A series of diodes that emit light in the infrared portion of the spectrum are embedded along the outer rim on two sides of the plate. These infrared light beams experience total internal reflection from the top surface plate, and are detected by light sensors that line the opposite two sides of the plate. When one touches the top plate, the reflected light beam is disturbed and the change in the light's trajectory is noted by the sensors. With appropriate software, the system can ascertain which location or locations caused the alteration in the light's motion. Unlike the distributed capacitor system, this technique is sensitive to anything that would change the light's reflection, such as contact with a nonconducting gloved finger or a stylus pen.

Using your phone, you open your airline's app and call up your boarding pass, represented digitally with a quick-response (QR) symbol. Luckily, you have been selected for TSA PreCheck. The laser scanner at the TSA agent's station reads the symbol on your phone and beeps to let you through, after the agent has compared your likeness on your driver's license with the real article. You place your luggage and briefcase on the conveyor belt that passes through the x-ray machine. You then join the line to walk through the **metal detector.**

I f you want to make a metal detector, first start with your electric toothbrush and its recharger cup. In order to recharge the battery in your electric toothbrush, you placed the handle into a plastic holder that was connected to the electric outlet. In the holder was a coil that carried an alternating current from the outlet that created a changing magnetic field. This changing magnetic field penetrated the second coil in the toothbrush handle, and the changing magnetic flux through this coil induced a current. The presence of any metal between the two coils will alter the amount of magnetic field passing through the second coil.

The handheld metal detectors used by those seeking spare change or buried treasure at the beach also use two coils to detect the presence of any metal. These personal metal detectors have a large disc at the end of a stick, and to search for metallic objects, you move the plate near the ground. The plate has two concentric coils—an outer coil around the rim of the plate that, through an alternating current, creates an alternating electromagnetic field (equivalent to very, very low-frequency radio waves); and an inner coil closer to the plate's center that detects any reflected waves. If the radio waves generated by the outer coil impinge on a metallic object, this induces an alternating current in the object that generates its own varying magnetic field in the opposite direction of the incident signal. The inner coil then detects this reflected wave and sends an alarm (such as a flashing light or beep) to indicate that the plate has passed over a good conductor of electricity.

The walk-through detectors at civic buildings and airports use a minor variation on this process. They run a current through a coil in one side of the arch and then abruptly turn the current off. The current in the coil creates a magnetic field, and when it suddenly stops, the field passing through the coil similarly disappears. But changing the magnetic field passing through a coil will create a current that briefly tries to keep the magnetic field unchanged. That is, there is a current induced in the main coil, created when the magnetic field rapidly decreases. In old-style incandescent lightbulbs, you can see this back-current when you switch off the light using a wall switch. Watch closely, and you'll see the light dim, very briefly grow brighter, and then be extinguished as the lightbulb turns off. The momentary relighting of the bulb is the induced current, with the circuit trying to prevent the change in the magnetic field created when the power is turned off.

The airport metal detector continually pulses a current through its coil in the archway and monitors the magnitude of this induced current. A metallic object that passes through the arch acts momentarily as an antenna, and when the magnetic field is turned off, the object also briefly generates a current, responding to the change in the magnetic field. The induced current detected in the coil in the arch increases slightly when the metal enters the detector and then decreases as the metal leaves. The detector typically pulses the current in its coil a hundred times a second and adds up all the induced currents to identify the presence of metal.

What if you forgot to remove a metallic object, and have to be checked using a handheld-wand metal detector? This device operates in a manner similar to the beachcombing metal detector: one coil generates a very low-frequency and low-power radio wave, while another measures changes in this signal due to any metal object.

The person ahead of you in line for the metal detector has really thrown a wrench into the works. The TSA PreCheck *detector should not have been set off by his wristwatch. The detector's sensitivity levels need to be adjusted, in order to avoid additional false positives. The TSA agent instructs you to use the adjacent **full-body scanner.** Emptying your pockets of everything, including the paper money you secured minutes ago from the ATM, you place it all in a small plastic dish on the x-ray machine's conveyor belt. You step into the scanner. With your elbows bent, you raise your arms above your head and, in a misplaced bit of vanity, suck in your belly. You stand still while the scanner bar orbits about you.* I surrender, *you say to yourself.* Just please don't let me miss my flight.

M ost scanners at airports use "millimeter waves" to check for any contraband underneath your clothing, using the principle of "backscattering." Instead of measuring the light that passes through an object, in what is called transmission mode, backscattering monitors the light scattered at the interface between changes in density. Your dentist uses the transmission method when she takes an x-ray of your tooth. The detector photographic film is placed behind your tooth, and the x-ray source is located in front of the tooth. The film thus records any x-rays that are transmitted through the tooth, and any change in density (such as a cavity or a filling) alters the amount of x-rays that can make it to the detector, showing up as a change in recorded signal intensity.

Another way to create an image is to measure the reflected or scattered waves whenever the medium that the waves encounter changes. Physicists refer to these reflected waves as "backscattered waves," and the scanners at the airport have two vertical rods that contain both a source and a detector of millimeter waves, one moving in front of your body and the other passing behind you, forming a complete scan.

A millimeter wave is an electromagnetic wave with a wavelength of roughly 1 millimeter (about 0.04 inch, which is over ten times the diameter of an average human hair). Another term for a millimeter wave is a "microwave,"* the same range of the electromagnetic spectrum that your cell phone uses for communication. Light with a wavelength of one meter (a little over three feet) or longer is used for radio and television, while wavelengths of one nanometer (one-billionth of a meter) are x-rays. Visible light has a wavelength between 400 and 700 nanometers, depending on its color.

The reason the full-body scanner uses microwaves with a wavelength of one millimeter is that electromagnetic waves of this wavelength can pass easily through your clothing but are reflected from your skin. The fibers in your clothing don't perturb the millimeter waves, as they

---

* Technically a wavelength of one millimeter is between the microwave and infrared regions of the spectrum, but it's closer to the microwave.

are small compared to the wavelength of the microwaves, just as a large ocean wave is not deflected when it passes by a buoy or a swimmer. Your skin, on the other hand, provides a large barrier to the microwaves, which are scattered back when they reach you. Any object secured under your clothing will reflect the microwaves differently than your epidermis, and this difference in backscattering is detected by the scanner. Just as you have to remove any metal objects from your person when passing through a metal detector, you have to empty your pockets and remove your belt when using the full-body scanner, so as to not obscure the imaging under your clothes.

No matter how long you stand in the scanner booth, however, you won't be cooked like a baked potato. The microwaves in the full-body scanner are much weaker than those used in your microwave oven, or even those used when sending a message to your cell phone.

While the millimeter wave scanner indicates whether you have anything hidden underneath your clothing, it provides a low-resolution image of any contraband. Though much less common than the millimeter-wave-based scanners, some airports have x-ray-based backscattering systems that provide a more precise image of what a passenger could be smuggling.

Speaking of x-rays, right now your carry-on bag is being examined using a solid-state method for imaging that does away with photographic film.

*While you stand in the full-body scanner, your carry-on luggage, along with the contents of your pockets, has passed on the conveyor belt through the x-ray machine. From your vantage point in the glass booth of the scanner, you can see the flat-panel monitor of an **x-ray scanner** at another security checkpoint. When a bag passes through the scanner, you see an image showing the outline of the bag and the silhouettes of various items inside it. You know that x-rays don't show colors the way we see them, yet the various items in the bag appear to be of several different colors. Despite the delay due to the metal detector and full-body scanner, your bags have not reached the other side of the security checkpoint yet. A TSA agent is having some difficulty making out something in your bag, and he has stopped and reversed the conveyor belt.*

The x-ray imager used by the TSA operates in transmission mode, exposing the top of your bag to x-rays and measuring what passes through. Your luggage passes over a solid-state detector plate that is divided into a large number of individual semiconductor pixels. We can think about nonconductors such as insulators or semiconductors as an old-style movie theater, in which all seats in the lower-energy orchestra are occupied by electrons and all seats in the higher-energy balcony are empty.*

---

* This is one of my favorite analogies for the band structure of solids, and I've used it many times. Perhaps I've spent too much time at the movies, and not enough in the library.

Only electrons excited up into the balcony, by either heat or light, will be able to move from seat to seat, carrying an electrical current. (Electrons don't like to be in the same location as other electrons, so they can only move over to another seat if it is empty.)*

The now-vacant seats in the orchestra (called "holes") can also move in response to an electric field—as adjacent electrons slide over, the empty seat acts like a positive charge moving through the solid. In an insulator the energy separation between the top of the filled lower band and the bottom of the empty upper band is in the ultraviolet part of the spectrum, which is why many insulators, such as window glass, are transparent to visible light. In glass, visible light does not have enough energy to bridge the orchestra and the balcony; this light cannot be absorbed, and it passes right through the material. Semiconductors, on the other hand, have an energy separation in the infrared or visible part of the electromagnetic spectrum, and thus it is easier to promote electrons into the higher-energy balcony states.

In some detectors the x-rays are absorbed directly by the semiconductor, while in other systems the x-rays first strike a material called a "scintillator," which emits a flash of visible light when exposed to high-energy radiation. The absorption of either x-ray or visible-light

---

* The formal name associated with this phenomenon is the "Pauli exclusion principle."

photons generates mobile electrons and an equal number of positively charged holes in the semiconductor detector. A small built-in electric field pulls the charges to the top and bottom of the device, where they pile up. The more light that is absorbed, the more charges that pile up, and measuring these charges provides a record of the light striking that particular pixel.

Before the x-rays reach the detector, they have to interact with the contents of the bag in the scanner. When trying to image the contents of a piece of luggage, contrast is king. For a given atom, the more electrons it has, the greater the likelihood that it will scatter x-rays from their initial trajectory. The baggage scanner uses x-rays with a range of energies (and hence wavelengths) in order to enhance the scattering sensitivity. Explosive materials tend to be composed of long-chain carbon molecules (remember that each carbon atom has only six electrons), and their atoms interact better with lower-energy x-rays. The steel in knives and handguns is mostly made of iron, which is more effective at scattering higher-energy x-rays. Your bag is exposed to both high- and low-energy x-rays, and two detectors with a thin layer of metal between them record what passes through—the top detector gets the full signal and the second detector is underneath a metal plate that filters out the low-energy x-rays. The intensity of the transmitted x-ray signal is color coded, depending on the x-ray's energy. By comparing the signal between both the high- and low-energy x-rays from the top detector and then looking at just the high-energy x-rays from the bottom detector, one can extract how many low-energy x-rays were scattered, improving the imaging of any organic material.

Your laptop is scanned separately, but not necessarily because the metal in the casing would shield any objects underneath the computer from the probing x-rays. Most cases are composed of either plastic (an organic material) or aluminum, which is one of the few metals with a low number of electrons (thirteen) per atom. The separate scan is because the computer interiors are dense and complex. There are so many small components within a laptop or tablet that the device warrants its own careful examination to make sure no contraband resides within.

Your luck seems to have run out. Your bag has caught the attention of the TSA agent operating one of the x-ray scanners. As you return the contents of your pockets back where they belong, you hear the four words you never want to hear at an airport security checkpoint: "Is this your bag?" You nod, and the agent tells you he wants to test it further. He takes your bag to a stainless-steel table behind the checkpoint and wipes the outside of the bag with a small white paper disc.

*He places this disc into a large, box-shaped device that you note is labeled an **explosive-trace detector**. Less than a minute later, the device gives the all-clear, and the TSA agent thanks you for your cooperation. You are finally free to go, and not a moment too soon. You will have to hurry to get to your gate on time.*

When a small paper disc is wiped over your luggage, any traces of the chemicals that are employed in explosives are transferred to the disc. If you had recently touched explosive powder, minute amounts would be embedded in the grease of your fingerprints. When you touched your bag, about a milligram of material could have been transferred. The explosive-trace detector is designed to "sniff" for specific chemicals on the paper disc.

We need some way to distinguish one type of molecule from another. A molecule of trinitrotoluene (TNT) consists of seven carbon atoms, six oxygen atoms, five hydrogen atoms, and three nitrogen atoms, all arranged in a very particular configuration. There are certainly larger molecules, but a molecule of TNT is many times bigger than the oxygen and nitrogen molecules that make up the atmosphere, and much more massive. Ideally we would like some sort of test that is sensitive to both the shape and the mass of the molecule so that it could pick out TNT from other organic molecules, avoiding the problem of false positives. One such method used to select the TNT molecules from the crowd is called "ion mobility spectroscopy,"* and it involves having all the molecules coming off the white disc run an "auto race."

Imagine that I install the same engine into two different vehicles, such as a large van and a small sports car, and I start them from the same point and have them race in a straight line—which vehicle would win? Because they have the same engine, the difference in travel time will depend (for the most part) on the difference in weight and air drag. The heavier the vehicle, the smaller the acceleration the engine can provide. Moreover, because the van has a much larger surface area and front profile, it will have greater air resistance. Any energy from the engine that is required to push the air out of the way as it travels down the racetrack is not available to increase the auto's speed. Therefore, I can distinguish between the van and the sports car by seeing which one crosses the finish line first, as long as they have the same engine, since the travel time depends on their mass and air drag differences. While we can tell these two vehicles apart by looking at them, the same process works just as

---

* Technically the spectrometer is detecting special "tagging molecules" that by law must be added to all explosives, but for simplicity we'll just refer to TNT.

well when we can't visually inspect the racers as when we are dealing with molecules.

The white disc that has been swiped over the luggage is placed at one end of a cylinder (the starting line of the track) and is then heated, driving the molecules into the vapor phase. We want to pull the molecules down to the other end of the cylinder (the finish line of the track), and we want the pulling force to be the same for all molecules, regardless of their chemical composition. One way to accomplish this is to remove one electron from each molecule, leaving them all with a net positive charge of +1. A negative voltage is applied at the finish-line end of the cylinder, and this pulls all molecules with the same force, as the only factor that governs how an object responds to a voltage is its net electrical charge of +1.

Having all the molecules have the same charge is equivalent to putting the same engine in all the vehicles in this race. The cylinder is filled with a particular pressure of air, to control the effects of air drag as the molecules move down the cylinder. Now, because they all experience the same pulling force from the negative voltage at the end of the cylinder, we note the time they need to reach the other end of the tube. The result is a spectrum of molecules separated by mass and size, arising from the different mobility of the ions, and so the explosive-trace testing station is an "ion mobility spectrometer." From prior calibration experiments, we know how long a molecule with the mass and structural configuration of TNT will take to go from one end to the other. If we record any molecules "crossing the finish line" at that particular time, then we have good reason to suspect that the bag in question was in contact with TNT at some point, warranting additional scrutiny.

CHAPTER FIVE

You Take a Flight

*Somehow, you've made it to your gate just as the second-to-last passenger is boarding. You call out to the boarding agent to hold the door. You run up, out of breath, and see that they are waiting for you! You apologize and quickly re-load your boarding pass QR code onto your smartphone. The laser scanner beeps, welcoming you to board. However, at this stage all of the overhead storage space is completely full. Your bag will have to be gate-checked. Not ideal, but much better than missing the flight. The gate agent asks if there are any electronics or a **lithium battery** in your now no-longer-carry-on bag. You inform her that there is not, and she attaches a red tag to your bag and asks you to take it to* *the end of the Jetway, right by the entrance to the aircraft. The ground crew, taking the last of the bags for storage in the cargo hold, accepts yours from your grateful hands.*

The flight attendant asked if there are any lithium batteries in your checked luggage for two reasons: (1) all batteries have the potential to release their stored energy in an uncontrolled manner if damaged, and (2) the high-energy density of lithium-ion batteries makes them particularly hazardous.

Lithium, like its alkaline cousins sodium and potassium, is a very chemically active metal, and in its pure, elemental state it reacts violently with water. As conventional alkaline batteries age, hydrogen gas can build up in the cell, exerting increasing pressure until the seal of the battery is breached. The alkaline metal will then react with carbon dioxide in the air, forming a furry coating over the electrodes of the metal terminal. As bad as this is for conventional alkaline batteries, lithium-based batteries been known to have more dramatic failures, resulting in fires or explosions. Their greater energy density means that much more energy is released if there is a short to the terminal separator, compared to a conventional alkaline battery. This is why one must take care when disposing of any battery, particularly a lithium-ion battery. Any puncture or crushing of the battery container can result in a dangerous short-circuit situation.

Lithium-based batteries are found in many personal electronic devices, as they have a better available-energy-per-pound compared to other batteries. The stored electrical energy in the battery comes from the energy released by the chemical reactions that pile up charged ions on the terminals. One way to get more charges in the electrodes is to use ions with more than either one extra or one missing electron. However, once one electron is removed from an atom, making it an ion, the now unbalanced positive charge of the nucleus has a stronger pull on the remaining electrons, and it becomes much harder to extract additional electrons or to add electrons to negative ions. If one can't easily vary the charge of the ions, another way to increase the charge-stored-per-pound of a battery is to use lighter ions, decreasing the weight of the battery. The lightest atom that is not a gas at room temperature is lithium, and lithium-ion batteries have played an important role in the development of portable electronic devices.

Years ago, batteries in automobiles consisted of lead-acid mixtures. Lead is a big, heavy atom, weighing over two hundred times more than a hydrogen atom. Hydrogen, the universe's lightest element, consists of a single proton in the nucleus orbited by one electron; lead has 82 protons and 125 neutrons in its nucleus, surrounded by 82 electrons. Protons and neutrons are roughly 2,000 times more massive than electrons, so to reduce the weight of an atom, consider elements with very light nuclei. Helium has only two protons and two neutrons in its nucleus, and is the second-lightest element after hydrogen, but it's a gas at room temperature and pressure, as well as chemically inert, and thus not suitable for electrical energy storage in a battery. The third-lightest element, with only three protons and three neutrons in its nucleus, is the alkaline metal lithium. This is the source of the big savings in weight of lithium-ion batteries.

Regardless of the type of conducting electrodes they employ, or the specific composition of the fluid in which they reside, batteries rely on a chemical reaction to generate ions in order to provide a voltage. This reaction expends energy, which forces the ions onto the terminals. In lithium batteries the terminals are composed of layered materials such as graphite or cobalt oxide that consists of stacks of atoms, like the levels in a parking garage. The lithium ions thus reside inside these terminals, parked between the atomic planes, enabling more ions to be stored. The maximum voltage that a battery can hold is determined by how much energy is available from the chemical reaction, as the more ions are added to a terminal, the harder it gets to force an additional ion into place. The positive ions on one terminal are electrically attracted to the negative charges on the other. All batteries must therefore have a barrier of some sort to keep the charged terminals separate, preventing the ions from recombining. This barrier must be porous, in order to allow the chemical reactants to move to the respective terminals, but it can't be such an open border that the negative and positive charges can leave their electrodes and recombine within the battery itself. If this barrier is breached, the battery can short-circuit, and the energy expended to charge up the terminal is rapidly recovered,* typically in the form of heat and gas due to the reversal of the chemical reaction.

You board the aircraft and make your way down the aisle to your assigned seat. It's not hard to spot the only unoc-cupied one in the full plane, way in the back, a window seat. You place your small bag underneath the seat in front of you and fasten your seatbelt. The main flight attendant an-nounces that the airplane door has been closed and that all personal electronic devices should be turned off or switched to airplane mode (you do the latter for your smartphone and check that your tablet is off). After a few minutes the plane begins taxiing toward the runway. The pilot comes on the public address system, welcomes you onboard, and indi-cates that your flight is number two for takeoff. The flight attendant begins the well-practiced safety presentation, and

---

* The energy recovered is actually less than the energy expended charging the terminals, due to thermodynamic considerations, but this technicality is small comfort if your battery explodes.

*almost as soon as she says the last word, the pilot tells the crew to prepare for takeoff. The plane idles for a moment at the start of the runway, and then the throttle is opened and the craft begins its acceleration. Gaining speed, the **airplane** begins, gently at first, to angle up toward the sky. Soon you are airborne.*

I f you want to leave the surface of the Earth, whether in a hot air balloon or a supersonic jet, you need more air molecules to strike the bottom of your craft than the top, leading to a net unbalanced upward force, and if this force is greater than your weight, it will lift you up, up, and away.

Hot air balloons and blimps rise for the same reason ice cubes float at the top of your drink rather than sit on the bottom of the glass—a difference in density leads to an upward force, called "buoyancy." The magnitude of the buoyant force depends on the size of the object, that is, how much of the fluid or air is displaced by the volume of the object. If the weight of the object is greater than the weight of the medium it displaces, it will sink; if it is lighter, it will float.* To get a balloon to rise, one fills it with a gas (typically either hot air or helium) that has less mass per volume (density) than the surround-

---

* Wood has many small pockets of air that lower its overall density, whereas when fully saturated with water, wood will sink to the bottom of a body of water.

ing atmosphere. For transportation in a device that has a density greater than air, one has to work a little harder in order to arrange for the atmosphere to provide a net upward force.

An airplane's wings are shaped and curved, with a larger profile on top of the wing, so that the air moving beneath the wing is deflected downward and the air moving above the wing turns upward. The air beneath the wing is jammed into a smaller volume, and when this happens the air pushes back on the wing in the opposite direction. Here we make use of Newton's third law of motion: *For every action there is an equal and opposite reaction*—or, forces come in pairs. You cannot push on something without that something pushing back on you. This deflection of the moving air creates a larger pressure (force per area) under the wing. Above the wing the air moves into a larger region, which decreases the pressure. The faster the plane moves, the more air per second is deflected, and the greater the pressure difference across the top and bottom of the wing. This difference in pressure caused by the effort of deflecting the flow of air leads to a net upward force. At a high enough speed, this upward force can be greater than the weight of the plane.

How does the plane move so fast that an upward lift is achieved? Once again, the operating principle is Newton's third law of actions and reactions. In the early days of flight (and still today, for smaller planes), the speed

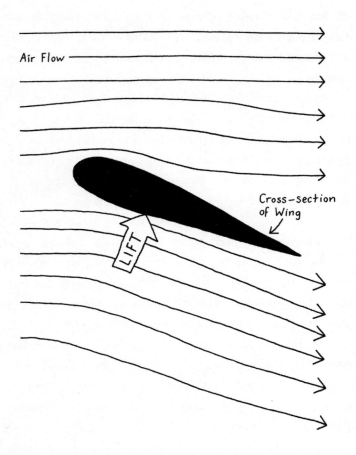

Air Flow

Cross-section
of Wing

LIFT

The air flow underneath the wing is deflected down.

The air pushes back with an upward force on the bottom of the wing.

This is called LIFT.

FIGURE 5

of the plane resulted from forward thrust obtained from rotating propellers. The blades of the propeller are shaped and mounted at an angle so that they are essentially rotating wings. They scoop and deflect a quantity of moving air behind the blade. This creates a pressure difference across the width of the propeller blade, pushing the propeller forward, in the direction of the plane's forward motion. The faster the propeller blades rotate, the more air they deflect behind them, and the faster the plane moves in the desired direction.

The larger the plane, the greater its weight, and for a given wing size, the faster it must move in order to achieve sufficient lifting force under the wings to raise the plane off the ground. The energy to rotate the propeller comes from the combustion of fossil fuel, essentially the same physics that leads to rotational kinetic energy in a car. Greater speeds, necessary for larger lifting forces, are achieved using a jet engine.

A jet engine may not look like it, but it is essentially an internal combustion cylinder, as found in a gasoline-fueled car, just turned on its side. Cold air is drawn into the front of the jet engine, via rotating turbine blades. This cold air is compressed, greatly increasing its temperature. The hot gas is now mixed with jet fuel, and a spark ignites the mixture, inducing combustion. The now very-much-hotter gas is expelled out the back of the engine, through a narrow vent that helps speed the hot gas even more. This ejection of hot gas toward the back of the plane provides a forward thrust, through Newton's third law. Unlike the cylinder in the internal combustion engine, these four steps occur continuously in a jet engine, pushing the large airplane forward at speeds great enough to enable upward lift.

For the airplane, the rotational speed of the propeller, or the forward thrust of the jet engine, must increase along with the plane's velocity, in order to counter air resistance. This drag is reduced at higher altitudes, where the density of air is lower. However, the thinner the atmosphere, the less air there is to push upward against the wings to provide lift, so one must still maintain a very high speed in order to fly at elevated altitudes.

*In an impressively short amount of time, your plane is passing through the clouds and climbing to its cruising altitude. You peer through the window and spend a few minutes quietly relaxing, letting go of the stress of the morning. The expanse of blue sky is calming, and you remind yourself that you are lucky to be living when you are—a hundred years ago, such vistas were not routinely available. You spot a striking cloud formation that extends for miles. You take out your smartphone and snap a **digital photograph** of the clouds.*

Two important steps are needed in order to photograph the image of the clouds using your phone:

recording incident photons via quantities of electric charge that remain localized at the positions (the pixels) where the photons are absorbed; and, once the photo has been taken, measuring how much electric charge sits at every position in the image. The first step is accomplished with a light-emitting diode run backward—that is, a light-absorbing diode, otherwise known as a solar cell or "photodiode." The second step involves arrays of capacitors that store these charges.

Sitting at each pixel is a photodiode that, through the absorption of light, generates electric charges. This device consists of a silicon chip sandwich, with pure silicon as the "meat" and silicon with intentionally added chemical impurities as the top and bottom pieces of "bread." The top slice has phosphorus added to the silicon, increasing the number of electrons in the normally empty upper "balcony" band of states. The lower slice has boron incorporated into the silicon, which pulls electrons out of the lower-energy band (which is typically completely filled), effectively adding holes to the layer. The net result of these impurity-added layers of silicon is the creation of an electric field across the pure silicon material in the center.

This electric field is important, for it will move the electric charges to where we want them. In the auditorium model of solid-state physics, if a photon of light strikes the cell with enough energy to promote an electron to the upper band of states, leaving a hole in the

lower band, then the electric field across the semiconductor will push these charges in opposite directions.
The electrons will pile up on one side of the "sandwich,"
and the holes will accumulate on the other. The more
photons that are absorbed, the more charges will be sent
to the opposite sides of the device. In pre-digital photographic film, light was absorbed by small grains of silver
halide that underwent a chemical reaction, permanently
altering the film. In a digital camera the charges from
the photodiode are stored in a reusable capacitor. At each
pixel the plates of a capacitor hold the electrons from the
photodiode at that location, and it is this charge that will
be read out in the next step.

Accounting for the charges stored on one million capacitors arranged in rows (where there are one thousand
rows with a thousand capacitors in each row), requires
careful timing and coordination. You want to make sure
that the charge determined to be on the capacitor in the
347th row and 589th column is assigned to that unique
location, all in under a second. The trick to accomplish-
ing this accurate and speedy read-out is the "bucket brigade principle."

In each row is a long line of capacitors, each with a
certain amount of stored charge. It's easy to read out the
charge on the first capacitor. A voltage is applied that
pulls this charge to a reference capacitor, just to the left
at the end of the row. The charge across the reference
capacitor induces a voltage that is detected and ampli-

fied using a transistor. The resulting voltage is converted into a digital signal using binary math, and this number is stored in a computer's memory. So now the capacitor adjacent to the end of the row on the left is empty, but we need to gain access to the charge on the next capacitor over.

In the bucket brigade method, the "pail" at the end of the line is emptied into a measuring barrel and then the pail next to it dumps its contents into the first pail. The third pail then sends its contents to the newly emptied second pail, the fourth pail transfers to the third pail, and so on. Instead of containing water, the capacitive buckets hold a quantity of electric charge, but the principle is the same. The positive voltage that emptied the first capacitor at the end of the row will attract the electrons that are held in the second capacitor until all of the electrons that were on the second capacitor have swung over to the first capacitor and then swung again, to the reference capacitor (where the charges are measured). The second capacitor is now empty, and the reference capacitor at the far left-hand end of the row is holding the charge from the second capacitor. This charge is sent down to the amplifier to be measured and converted to a voltage. Meanwhile, a positive voltage is applied to the plates of the now empty second capacitor, attracting the electrons that are on the third capacitor over. This process continues for all one thousand capacitors in the row, and for all one thousand rows, until a million capacitors have been examined and the charges they were storing have been converted to a set of one million voltages. This array of capacitors, read out in this manner, is technically termed a "charge-coupled device," or CCD.

The key to making this system work is coordination. In order to store the voltage on a memory chip in the proper sequence, so that the information encoded in the sequence of charges can be reconstructed into the original image, it is crucial that all of the capacitors transfer their charges down the line at the proper times. This depends on precise computer clocks that ensure correct timing.

The procedure described above will convert a color image into a black-and-white photo, albeit with a fine gray scale at every individual pixel. That is, the magnitude of the charge, and hence voltage recorded from each capacitor, depends on the number of photons striking that picture element's light sensor, and because this voltage can have any value, it can capture shades of gray. One technique in digital photography to create color photos involves dividing each sensor into four equal, smaller squares. Over each of these smaller squares is a filter— red, green, blue, or a clear filter that allows all the light onto the detector. The separate charges sent over to the amplifier at the end of the row keep track of which color's capacitor is being recorded. Processing, smoothing, and interpolating the voltages can adjust the color levels of the reassembled image. One is not limited to the visible portion of the spectrum. Semiconductor sensors, with much smaller energy gaps between the filled lower band and the empty upper band of energy states, can detect infrared light. As in the case of film photography, once a photon is absorbed and converted into an electronic signal, the rest of the processing is independent of the wavelength of the light that was detected.

The development of CCD-enabled cameras has revolutionized photography. The light from distant galaxies, or the view from a jet plane flying at over 500 miles per hour, can be accurately recorded using the same basic technique that Mathew Brady employed in his Civil War–era portraits when he left the camera's aperture open and waited for enough photons from his subject to reach the camera's detector. Brady's subjects had to remain completely still for half a minute in order to avoid blurring of the image, but modern cameras can take a photo in a fraction of a second.

The smaller the pixel size, the more picture elements are needed to cover a given area, and the higher the resolution. This holds equally for cameras and displays. We have used as an example a digital camera with one thousand rows containing one thousand sensors, but the million pixels (or one Megapixel) that this represents is near the low end of present-day resolution. Today, state-of-the-art resolution for digital cameras is roughly twenty Megapixels, but when it comes to technology, art never stays in the same state for long.

Once the information contained in the stream of

photons has been converted into voltages, these voltages can in turn be transformed into a sequence of ones and zeros, the precise sequence of which can encode instructions for which pixels on a display should be activated at which locations and times. If sufficient computer memory is available, then you can take multiple images—a new one every forty or fifty milliseconds—and this video image will be described by a much larger set of ones and zeros. It's these ones and zeros that you wish to upload onto a remote data storage center—you want to send your photo of a cloud to the Cloud.

*Your phone will automatically save any digital photos you take in your **cloud data storage** account. However, with your phone in airplane mode you have no access to your data provider. You will double-check when you land that the photo is saved alongside your other photos and videos. The plane is now well above the clouds, and you recall that with less atmosphere above you, you will be exposed to more radiation on this flight than during your doctor's visit today. This reminds you to look over your notes for your presentation.*

For many of us, the number of ones and zeros that we wish to maintain for long-term storage eventually exceeds the capacity of our smartphones. It thus becomes a better choice to keep our data in a "safe-deposit box" that has very deep drawers. The bank vault that contains all of these virtual, digital lockboxes is sometimes referred to as "the Cloud."

If you are up in an airplane, five miles above the ground, then you are roughly five miles farther away from the Cloud. The information that you store there resides on the ground, often in the globe's north. Unlike your local savings and loan, it does not have a definite central location; your information is preserved on a multitude of computer memory banks, referred to as "servers."

When you turn on your phone or tablet, your device runs an instruction set that tells your display which pixels to light up. Similarly, when you request, via a Web address, information such as a site (with text, links, and graphics), a different instruction set is sent to your device, which modifies the display accordingly. The website's information, the complex pattern of ones and zeros sent to your tablet, is stored on a server. The totality of the servers that preserve your information, in a diffuse and extended network, is referred to as a "cloud data storage" system. But care must be taken to make sure that all this data does not "fry" the servers.

There is a small but non-zero cost in energy every time a transistor switches from the low-current "off" state to the high-current "on" state. If that energy is not drawn away quickly enough, then the temperature of the transistor will increase. Temperature is a way of keeping track of the average energy per atom in an object. In a solid, the atoms are locked into fixed locations, and the higher the average energy, the more the atoms vibrate about their equilibrium positions. This shaking can deflect an electrical current away from its trajectory, limiting how large a current can flow for a given applied voltage. In addition to decreasing the performance of the transistor, this increased temperature can actually destroy the chip itself. If the heat is not removed, the buildup of energy can actually raise the temperature above the melting point of silicon, as many millions of transistors in a very small volume are switching millions of times a second.

Heat management is a major issue in the design of computer systems, particularly when one wants to maximize the speed at which the system operates. The faster the transistors switch, the shorter the time interval between changes in current level, and the less time available to remove the heat generated by this operation.

There are many different companies that provide cloud data storage, and each one uses a vast array of computer servers, all of which need electrical power. The data centers run by Google use over two hundred MegaWatts of power, the equivalent of the output of a small power plant. Multiply this by the large number of data centers around the world, and it is clear that cloud storage (and the Internet in general) has a considerable carbon footprint. In order to save on air-conditioning costs (if the servers overheat, then there go your saved cat videos), many of these data centers are situated in colder climates, such as Sweden, Finland, and northern Minnesota.

*Satisfied that you are ready for your presentation, you take your tablet out of your bag, along with a headset, intending to pass the remainder of the flight by watching a video you have downloaded. You are impressed by the quality of sound provided by the diminutive earbuds. You turn to the person sitting next to you and see that he is watching his own video on his tablet, but instead of small earbuds he is wearing **noise-canceling headphones**. These headphones are large and bulky, and you wonder whether they truly cancel the noise or just muffle it.*

The earpieces in noise-canceling headphones are large not only to provide insulation against background sound but also to allow each earpiece to do the job of a telephone handset. That is, they function both as a microphone—detecting incoming sound waves, converting them to electrical voltages—and as speakers that translate these voltages back into sound waves. In order to cancel the background noise, these new sound waves result in perfect destructive interference with the incoming sound.

Sound is manifested by density variations in a medium, typically air. These density modulations can be smoothly periodic, with a well-defined frequency, such as result when a single guitar string is plucked, or they can be random and chaotic, like the roar of the plane engine. Consider the note from a guitar string. As we have seen, when the string moves toward us, it collides with air molecules, compressing them with the air molecules in the space ahead of them, creating a region with a higher-than-average density. As the string then moves away from us, it leaves behind a space with a lower-than-average density that also propagates away from the string. When the string vibrates back toward us, it generates another higher-density packet of air, followed again by a lower-density region. The back-and-forth vibrations of the string continuously create alternating variations in the density of the air. The spacing between adjacent regions of denser air is called the "wavelength." Let's say the density wave resulting from the vibrating string is the note G. This wave beats a tattoo on our eardrum, causing the membrane to vibrate. Eventually this mechanical vibration is converted into an electrical signal that is transmitted and processed by our neurons, which interpret the incoming signal as the note G.

But now imagine that before this wave reaches our ear, it collides with another wave that represents the same note. Everything in the second wave is identical to the first, with one crucial difference: this second wave is perfectly out of phase with the first one, so that whenever the first wave has a region of higher density, the second wave has a region of lower density, and vice versa. In order to create this second wave, you just need another guitar string, which you pluck exactly as you did the first, but taking care that when the first string is moving down, the second string is moving up. When these two waves meet, the higher-than-average density regions fit snugly into the lower-than-average density regions, restoring the air to its average density at all points in space. The second wave effectively makes it as if the first wave didn't exist. Air with a uniform average density is just air—and the guitar note G disappears. The first guitar string created a sound wave, and the second string generated an anti-wave—canceling it out.

What can be done with one single note can be done with many different notes. The vibrations of the plane's engines are transmitted through the frame of the wing to the fuselage, and from there to the air in the pressurized cabin. These sound waves are distinct from music or conversation, as the plane's background noise has a broad range of frequencies, all competing for attention when they strike your eardrum. Imagine you are eavesdropping on a conversation across a large, empty room. With effort you can make out what is being said. But if the room is hosting a large party, with dozens and dozens of couples all having their own conversations, it would be much harder to continue to listen in on one particular discussion. Regular headphones will block some of the background noise while piping in the desired signal directly to your ear, but some of the noise will still get through, making it harder for you to concentrate on what the headphones are playing.

Noise-canceling headphones record the conversations of everyone else in the party, and instantly generate anti-waves that negate their sound waves. The microphone built into the headphones detects the incoming sounds, converts the density waves into corresponding voltage variations, and quickly analyzes the input. Once the noise has been analyzed, it is a straightforward mathematical procedure to flip the phase, creating an exact anti-noise voltage. This voltage is sent to speakers in the headphone that generate a mechanical sound wave canceling out the incoming noise wave. Since the speed of the electronic processors built into the headphones is many times faster than that of sound, the headphones can compensate when the noise changes, always creating a near-perfect cancelation.

Before your video is finished, the captain's voice comes on to announce that your flight has begun its descent. You are asked to put away any personal items you may have removed during the flight and turn off any electronic devices until the *plane has landed. You stow your tablet and earbuds, and soon enough you are on the ground. After deboarding the plane, you retrieve your luggage that was gate-checked. You join the crowd of people exiting the airport. The line for a taxi is way too long, so you decide to take the light-rail train to your downtown destination. Thinking about the quick and efficient light-rail system, you muse that had there been a **high-speed train** alternative to the flight, you'd have taken it.*

From a physics perspective, the essential issues for a light-rail train are basically the same as for an all-electric car. However, the fixed track provides yet another option for propulsion that can be employed in high-speed trains—magnets.

Why are some materials magnetic in the first place? The constituents of atoms—the protons, neutrons, and electrons—all have a small built-in magnetic field, like a compass needle, accounted for by quantum mechanics. In an MRI scanner, the intrinsic magnetic field of the atoms' nuclei enabled us to image the inside of your body. The magnetism of conventional magnets (bar magnets or horseshoe magnets) arises from the magnetic field of the electrons in iron atoms. While most atoms have their electrons form pairs, north pole to south pole, so that the two of them together have no net magnetic field, in a few cases (such as iron or cobalt) a number of unpaired electrons (though not all of them) have their north and south poles all pointing in the same direction, giving the atom a net magnetic field. The natural tendency of the iron atoms is to have their magnetic fields point in the same direction, and a solid piece of iron can be magnetized, creating a large magnetic field in the space around it. Bring two such magnets together, with their north poles facing each other, and there will be a strong force pushing them apart. In this repulsive configuration, if one magnet is on the railroad track, and the other is on the train, this repelling force can be greater than the weight of the train (provided the magnets are strong enough), which will levitate above the track.

This repulsion between similar magnetic poles is how magnetically levitating trains ("maglev," for short) float. But they don't use iron bar magnets; instead, maglev relies on powerful electromagnets. We have invoked the very deep connection between electricity and magnetism, exploited in the motor of your electric toothbrush and the transformer in its recharger. Moving electric charges create magnetic fields, and changing magnetic fields can induce electrical currents. In order to generate magnetic fields large enough to lift a train off the track, a very large electrical current is required.

The track that guides the maglev train consists of electromagnets that both levitate and propel the train. A coil that will generate a magnetic field is on the bottom of the train. If the current flows clockwise around the coil, then the north pole points out of one end of the coil, and if the current is made to flow in the counterclockwise direction, then a south magnetic pole comes out of the coil. The current in the coils in the electromagnet in front of the train is set up so that as it generates a north pole magnetic field, it is attracted to the region of the track with a south pole, pulling the train down the track. As soon as the coil is aligned with this south-pole region of the track, the current in the coil is reversed so that now it generates a south pole—pushing the train away from this track section and toward the next, which has a north pole. With a computer guiding the flipping electrical current, the changing current in the coil will continue to pull the train faster and faster down the track. Another set of coils lifts the train up off the track, so the friction between the rotating wheels and the track is eliminated. Magnetically levitated train speeds of over 200 miles per hour are possible.

On the light-rail train, you take out your smartphone and connect to your data provider. You verify that your photo of clouds has indeed been stored in your Cloud account. You then call up on a Web browser a map of the light-rail train you are riding and cross-check it with the downtown office building that is your destination. Six more stops to go. You have enough time to reflect on what you are going to say.

CHAPTER SIX

You Give a Business Presentation

You get off at your stop, two blocks from the office building where your meeting is being held. As in most American cities, downtown is laid out on a uniform rectangular grid, and once you figure out the correct direction, it's easy to find your destination. Arriving at the building, you check the directory and hop in an elevator, riding it to the twenty-third floor. After you introduce yourself to the receptionist, she telephones your host. A few minutes later you are led down a hallway to the meeting room where you will give your presentation. It's a fair-sized room, and your host indicates that over fifty people are expected to attend. Your talk is scheduled to begin in twenty minutes, and while you have a coffee, *you try to connect your tablet to the LCD projector. You have some trouble getting your tablet to communicate to the projector. Fortunately you have a backup copy of your presentation on a **USB flash drive,** which you use to load your talk onto a computer in the meeting room. After verifying that the file opens and is still properly formatted, you are ready to begin.*

Computers register ones and zeros using semiconductor transistors that represent these values through high and low current levels, respectively. The cool thing about semiconductors is that their electrical resistance can be dramatically changed either by adding tiny amounts of other chemical elements (the basis of solar-cell and light-emitting diodes) or by applying an electric field (in a transistor).

To make a transistor, start with a capacitor consisting of two metal plates with a thin insulator (such as glass) between them, and replace one of the metal plates with a semiconductor, such as silicon. In a regular capacitor, a positive voltage is applied to one plate, and the metal sheet acquires a positive charge. This large positive charge attracts electrons to the second plate, leaving this sheet with a negative charge. When the second plate is a semiconductor rather than a metal, the same thing happens. The electrons in the silicon are attracted to the side of the silicon facing the metal plate. Previously, this region of the semiconductor at the interface with the insulating glass had only a few spare electrons to carry a current, but now the density of charges on the surface is much higher, and a larger current results across this region.* By applying a voltage to the metal plate, we can dramatically increase the current that can flow through the semiconductor. When a larger current moves through it, that can represent a "one" in the computer memory, and when no voltage is applied, then there is a very small current through the silicon (a "zero" in this case). This phenomenon—in which changing the charges on the bottom metal plate sets up an electric field that changes the resistance of the semiconductor—is called the "field effect," and the entire device is referred to as a field effect transistor, or FET.

The trouble with this transistor structure is that the device will have a high current, and thus register a "one," only when a voltage is sent to the metal plate. The energy to supply this voltage in your computer comes from the battery, or from a power plant if the computer is connected to a wall outlet. As soon as the voltage is turned off, the excess charge on the metal plate flows away, and

---

* Of course, by making one side of the silicon a better conductor of electricity, we have done so at the expense of making the other side an even worse conductor. But when you try to pass a current through a material, it will flow where it is easiest (the "path of least resistance") and we will see a larger current through the silicon resulting from the voltage applied to the bottom metal plate.

the transistor reverts to its low-current state. So all of the information stored in the sequence of ones and zeros is lost, as all of the transistors default to the zero state, unless you have transferred the digital information to a more permanent storage medium, such as a magnetic hard drive.

A USB* memory stick maintains the information of the high-current state of the transistors, even when disconnected from a power supply. In order to keep the transistor in a high-current state, a third metal plate, buried within the insulating glass between the first metal and the semiconductor, is used to store the electrical charges needed to keep the silicon in the "on" state.

The insulating glass layer in a transistor is typically very thin, only about one hundred nanometers thick (one nanometer is one-billionth of a meter and is the length of three atoms laid end-to-end). For a transistor in a USB drive, another metal plate is added, identical to the bottom electrode but sandwiched in the insulating layer, only ten nanometers beneath the silicon. This metal is not connected to any outside wires, so ordinarily it plays no role in the operation of the transistor. But if the semiconductor's wires are shorted out and a large voltage is applied to the bottom metal electrode (sometimes called

---

* USB stands for "universal serial bus"—which describes the thin rectangular connector that plugs into your laptop or tablet. USB was set as a standard design by the computer industry so that different devices could be connected with a single type of cable.

the "gate electrode" because the voltage applied to it allows a large current to flow in the top semiconductor, like swinging open a gate), then some of the charges can effectively jump* through the insulator and embed themselves onto this buried metal plate. Since the buried plate is not connected to anything, these charges will stay on this plate until a compensating voltage is applied, canceling it out. Until that voltage erases them, the charges on the plate inside the insulating glass induce an oppositely charged conducting channel in the semiconductor. This "floating-gate field effect transistor" will maintain its highly conductive "on" state, even when the transistor is unplugged from a power supply.

USB memory drives (also known as a "thumb drives" or "junk drives"), as well as tablet computers and lightweight, ultra-thin laptops, employ this type of transistor for long-term data storage. Advances in device manufacture, similar to those that led to a dramatic increase in the number of transistors on a chip, have rapidly expanded the capacity of computer memories using these floating-gate transistors. Computer hard drives originally employed** discs with magnetic regions to represent the ones and zeros. These discs rotate at very high speeds, and a magnetic read head, playing the same role as a coil,

---

* The charges get to the buried plate through a quantum mechanical process termed "tunneling," but we don't need to know the details.
** And many still do.

would convert these changing magnetic fields into electrical currents. In contrast, floating-gate FET hard drives have no moving parts, and the energy requirements of this form of computer memory are much lower.

*Your host pops back into the meeting room while you are transferring your talk from the USB drive to the computer and asks if it would be possible to get a hard copy of your presentation. This would make it easier for the audience to follow along and take notes during the talk. You've brought along only one copy of printed-out notes and slides for your presentation, in the interest of not weighing down your luggage for an overnight trip. Assuring you that this is not a problem, your host takes your notes and makes copies for everyone at the meeting using the nearby **photocopier**.*

Prior to the invention of xerography, if you wanted a copy of a document, you had to generate it when the first document was created, using carbon paper. In a photocopier, the content of one sheet of paper is transferred to another through a complex dance of optics and electronics.

A photocopier starts with a semiconductor coated onto the outside of a cylindrical drum. A wire at a very high voltage is used to spray electrical charges (let's say positive charges, for the sake of example) on the semiconductor's surface. The document to be copied is placed on a glass surface above the semiconductor drum. A lamp moves underneath the document, shining a bright light along the page. Where there is a white space on the document, this light is reflected onto the electrically charged semiconductor surface; where there is black printing, no light is reflected.* Any region that is exposed to the reflected light has its electrical charge neutralized, so the dark regions on the original document are now accurately reproduced by the remaining positively charged regions on the semiconductor drum. Nonconducting hollow plastic particles filled with a pigment (called "toner") are negatively charged using another wire with a different voltage. The toner is then sprayed onto the semiconductor surface. The negatively charged toner particles (primarily clumps of carbon atoms, along with some other chemicals) are attracted to the positively charged regions on the drum surface that correspond to the black portions of the original document. A sheet of paper is positively charged, using the wire employed when charging the drum, and physically placed on the drum, sandwiching the toner between the semiconductor and the paper. The negatively charged toner particles are attracted to the positively charged paper, and when the paper is peeled off the drum, the toner will come along with it. The paper is then heated, melting the toner

---

* For simplicity, we'll restrict ourselves to black-and-white copying.

onto the paper, and the electrical charges are neutralized. The paper is now a copy of the original document.

As the paper is ejected to the out-tray, a squeegee runs over the surface of the semiconductor drum, swiping away any leftover toner. This also cancels any remnant electrical charges on the surface, preparing the drum for another copy. All this happens, in most photocopiers, in less than a second.*

The key to making all this work is that when light shines on the positively charged semiconducting drum, the charges are conducted away, leaving only those sections in the dark with a positive charge to attract the negatively charged toner particles. Light can change the charge of the semiconductor surface because semiconductors are middling conductors of electricity—not quite as good as metals but much better than insulators. The seats in the lower-energy orchestra are all occupied with electrons, and because no two electrons can sit in the same seat, no current can move if there are no empty seats for the electrons. When light of the proper wavelength shines on a semiconductor, its electrical resistance can be dramatically reduced. Some of the electrons are kicked up into the higher-energy balcony states, and are able to move from seat to seat in response to an applied voltage, forming an electrical current. When visi-

---

* To quote David Owen, in a 1986 article: the Xerox machine "is a marvel of extremely intricate engineering that, like the Post Office, actually works much better than one has any right to expect."

ble light shines on the positively charged semiconductor, the conductivity of the illuminated region increases. The charges in the bright regions can rapidly move away to an electrode on the opposite side. Charges in the dark cannot move (their resistance is still too high), and so they remain to attract the negatively charged toner particles.

A similar mechanism also makes scanners and printers possible. Once the dark and bright portions of the document are mapped onto pixels on the semiconductor drum, the distinct regions of high and low charges can be stored in capacitors at each location. Using charge-coupled devices (think of a digital camera), they can be read out as a series of voltages and stored in any computer memory. A hard copy can then be created using this set of voltages to guide an ink-jet printer. In this device, a sheet of paper is pulled past a bar, where a belt and pulley manipulate a printing cartridge back and forth. One form of ink-jet printing uses a reservoir of ink, connected to a cylinder that has a nozzle through which ink can be sprayed onto the paper. The ink is ejected when pressure is applied to the top of the cylinder, where a piezoelectric crystal sits. As the voltage across this crystal is changed, the crystal either expands, pushing ink out through the nozzle, or contracts, drawing more ink from the reservoir into the printing cylinder. Converting documents into a series of voltages, and these voltages back into a printed copy, is why a single device can function as a scanner, a fax machine, a printer, and a copier: it's all the same physics!

For such a widely used product, the disordered semi-conductors essential to a photocopier are actually very poorly understood. Physicists have a much deeper understanding of the properties of crystals than they do of disordered solids. A cubic inch of any given solid contains nearly a trillion trillion atoms. In order to avoid the headache of solving a trillion trillion equations, physicists exploit the symmetry of the crystal structure to elucidate the properties of the material—but it's impossible to use this shortcut when the atoms are in a random configuration.

The semiconductors discussed so far are "crystalline semiconductors." Think of a solid composed entirely of carbon atoms, but with every atom sitting at a very precise location: at the center of a pyramid* with other carbon atoms at the points. Such a configuration is called the "diamond crystal lattice," and in fact, what we call "diamond" is carbon atoms in just such an arrangement. There are different types of crystal structures,** such as the structure of table salt, which is a cube, with sodium and chlorine atoms on opposite corners of the cube, repeated indefinitely. But there are many, many more ways

---

* Technically termed a "tetrahedron."

** The number of unique possible crystal structures is, however, surprisingly low.

that atoms can be arranged randomly, as in a disordered or amorphous structure. The semiconductor drum, a vital part of the copier, is in fact made up of an amorphous semiconductor (different from the crystalline type).

The drum or belt in a photocopier utilizes an amorphous semiconductor to keep costs down. To make a large, single crystal of silicon, you start with a vat of molten silicon. A crystalline seed is dipped into this high-temperature liquid and slowly retracted, and the silicon that is withdrawn adopts the seed's crystal structure as it is raised out of the molten pool. At the end of this slow and careful process there is a large cylinder of crystalline silicon, with a diameter of ten to twelve inches. The cylinder is then sliced into thin discs, like cuts of salami. (To use the entire crystalline cylinder for just one drum in a photocopier would make these machines so expensive that we would all still be using carbon paper.) An array of microchips is then fabricated on each disc through a sequence of chemical and thermal processing steps, and then the disc is cut into distinct chips that are employed in cell phones, laptop computers, and other technological devices. But for making copies on $8\frac{1}{2}"$ × $11"$ sheets, a disc is not what is needed. When the first copy machines were introduced, the largest cylinder of single-crystal silicon that one could manufacture was only two to three inches in diameter, insufficient for a copier. Cutting and gluing together pieces from several discs would leave unmistakable vertical streaks in the copy along each crystal-section seam. However, if atoms were essentially randomly deposited onto a surface, an amorphous semiconductor could cover an area the width of a standard sheet of paper.

Photocopiers, from the very first Xerox-brand photocopier, have always used amorphous semiconductors for the drum material. Despite their being employed in other technologies—such as solar cells, flat panel displays, scanners and copiers, and x-ray machines (basically, anything that requires semiconductors to be spread out over large flat areas)—scientists have an incomplete picture of how they work. Research continues, with the hope that a better understanding will result in better devices.

*Your host returns with the copies, and you connect the cable from the computer you are borrowing to the **LCD projector** in the meeting room. A moment of anxiety as to whether this computer will work is allayed when the desktop is projected on the presentation screen.*

A simple slide projector shines a bright light through a small photograph printed on transparent plastic, whereas an LCD (liquid crystal display) projector generates, from a digital file on your computer, a *virtual* photographic slide that can be projected onto a screen.

A liquid crystal is a true fluid, with all of a liquid's characteristic properties: it fills up the volume of its container, and when the container is tilted, the collection of molecules flows and pours like any other viscous liquid. But there is also an underlying order to the arrangement of the liquid crystal's molecular units—as one would find in a crystal. Most crystalline solids have single atoms or small molecules as the basic elements periodically repeated throughout the material; in liquid crystals this role is filled by larger, carbon-based molecules. The properties of these molecules are very sensitive to their detailed chemical components and their bonding arrangements. Depending on the precise composition and how the molecules interact, they can be randomly arranged as in a normal liquid, but when the temperature is lowered they may adopt a less random configuration.

Unlike the melting of ice or the boiling of water, the transitions that liquid crystals undergo involve less dramatic rearrangements of the molecules. For example, certain long-chain molecules, due to electrostatic interactions with their neighbors, can arrange themselves so that they all line up pointing essentially the same way, like a large number of toothpicks in a long narrow box (this is called the "nematic" phase). Lower the temperature, and the molecules now have even less thermal (that is, kinetic) energy. They are thus susceptible to another weak electrostatic interaction and line up in another pattern. The beauty and power of liquid crystals is that, for certain molecules, this transition from one molecular arrangement to another can be achieved by applying an external electric field across the cell in which the liquid resides.

It is this feature of liquid crystals that is exploited in LCD projectors, as well as in flat panel displays like tablets or televisions. The screen is divided into a two-dimensional grid of picture elements (like the ATM touch screen), where each pixel is a small cell consisting of sheets of glass held a small distance apart, and between the sheets is the liquid crystal. The front and back sheets of glass have polarizers that will only pass light whose electric field is oriented in a given direction. Ordinary white light is a jumbled mix of electromagnetic waves with electric fields oscillating in all possible directions. The polarizer is like the bars in a prison window, and once light passes through it, the electric fields will point in the same direction as these nonconducting bars. One end of the pixel cell has a vertical polarizer, so the light entering the cell will have oscillating electric fields in the vertical direction. The second glass plate has its polarizer oriented horizontally, and as the electric fields of the light do not lie in the horizontal direction, none of the light that makes it through the first glass is able

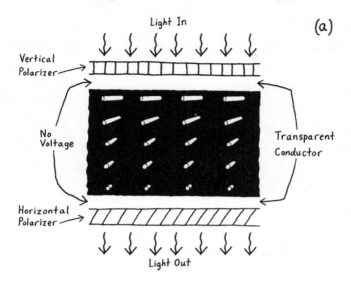

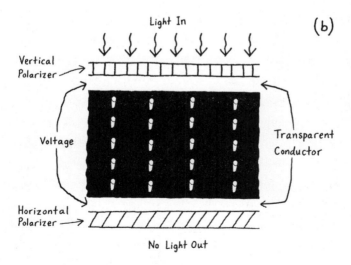

FIGURE 6.1

to pass through the second one—that is, unless the liquid crystal between the two polarizers can reorient the light's electric field.

A particular liquid crystal molecule is chosen for the LCD pixel, so that the long axis of the molecule lines up in the direction of the vertical polarizer's lines on one glass plate. The molecules on the other side of the plate align with the horizontal polarizer orientation. Between these two plates, the liquid crystal molecules undergo a gentle twist from the vertical to the horizontal alignment. The liquid crystal guides the electric field of the light, as it follows the orientation of the molecules. The electric field of the vertically polarized light rotates so that it is in the horizontal plane when it reaches the second polarizer. So this pixel will be bright when light shines through it.

If a voltage as low as five Volts is applied across the two glass plates (using transparent conductors such as indium tin oxide, as in touch screen displays), then the liquid crystal undergoes a transition into a different configuration. The electric field of the light now does not rotate from the vertical to the horizontal, and thus no light is able to pass through the now dark pixel.* By changing the voltage applied to each pixel in the array and using filters, we can generate a color image.

---

* Because the light emitted from a liquid crystal display is polarized, when you view such displays while wearing polarizing sunglasses, you can see interference patterns due to the different polarizers.

*Your first slide is projected on the screen, and you test out the small, handheld **laser pointer** that you'll use to indicate various features and bullet points in your presentation. The bright green dot is clearly visible even at the back of the room.*

Laser pointers, bar-code readers, and DVD and Blu-ray players all employ a device called a "laser diode," a refinement of a light-emitting diode.

A diode consists of two semiconductors placed together, with electrical leads arranged so that the current must pass through first one material and then the other. Recall the analogy of semiconductors as an auditorium with a filled lower-level orchestra, every seat occupied by an electron, and an empty upper-level balcony. Electrons can carry a current only if they move from one seat to an empty adjacent seat. The filled orchestra cannot carry a current, as there are no empty seats for the electrons to move into; nor can the balcony, for while it has plenty of empty seats it has no electrons. Imagine a semiconductor in which a small amount of a certain chemical impurity has been added, with the effect that the impurity donates electrons to the balcony, without taking any out of the lower-energy orchestra. This material will now be a much better conductor of electricity.

There are other, different chemical impurities that could grab electrons out of the orchestra, leaving empty

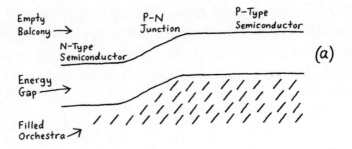

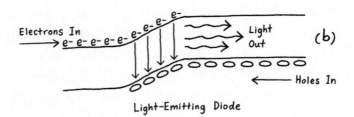

Light-Emitting Diode

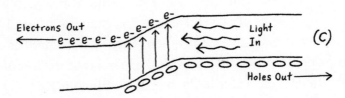

Photodiode, or Solar Cell

FIGURE 6.2

seats behind, without adding any electrons to the balcony. As an electron in the orchestra jumps into a now empty seat, it leaves an unoccupied space behind, and the missing electron (or "hole") can move through the orchestra, carrying a current. The first type of impurity would yield what is referred to as an "n-type semiconductor," so named because the mobile charges are *negatively* charged electrons. The second type of impurity creates a "p-type semiconductor," since the holes in the orchestra act like *positively* charged particles. A diode is what you get when an n-type semiconductor is placed next to a p-type semiconductor, and metal leads are attached to the n-type and p-type sides.

When a semiconductor with a lot of electrons but no open seats in the orchestra (n-type) is placed next to another material with plenty of empty orchestra seats but no electrons in the balcony (p-type), the extra electrons from the balcony on one side move over to the balcony on the other side (for the same reason that people in a crowded room will exit when the door is unlocked to an adjoining empty room), and this places them above the empty orchestra seats in the p-type material, allowing them to lower their energy by dropping down into the unoccupied positions, emitting energy as light or heat in the process. Most of this recombination takes place at the p–n junction, where the two materials meet. But the removal of the excess electrons and holes leaves behind other charges. The impurity atoms added to the two semiconductors have their own electrical charges (positively charged ions in the n-type semiconductor and negatively charged ions on the p-type side), and without the electrons or holes to balance them, they create an internal electric field. This field makes the diode a surprisingly versatile device—it's very hard for a current to flow against the internal electric field (even though there are empty seats on the balcony available to them), but very easy, if the wires' polarity is switched, for the current to pass in the direction of the field.

Shine light on the semiconductor, and if the light's energy is absorbed (that is, it spans at least the top of the orchestra and the bottom of the balcony), then an electron and hole are created. These charges suddenly find themselves subject to the built-in electric field, and are pushed apart, with the electrons moving to the n-type side and the holes to the p-type side. As more light is absorbed, more charges are pushed apart, and a current can be generated from the p–n junction, just by shining light on it. This is a solar cell, a way to convert the energy of light directly into an electrical current, without having to rotate coils in magnetic fields. This device is also the photodiode that converted absorbed photons into electrical charges in your digital camera.

If you force a current to flow in the opposite direction, pushing electrons in on the n-type side and holes in from the p-type side, they will meet at the p–n junction between the two materials; when they do, the electrons will fall down into the empty seats (holes) in the orchestra, and release their energy as light. This is a light-emitting diode (LED), and it is nothing more than a solar cell run in reverse. That is, for a solar cell, light goes in and current comes out, while for an LED, current goes in and light comes out. Same device, just run forward or backward.

In an incandescent lightbulb, a large current is passed through a small wire, and the kinetic energy of the electrical current is transferred to the wire, making it white-hot. Just as for a toaster wire, a great deal of the current's energy goes into heating the wire, and only a fraction of it is converted into light. But in an LED, there is little heat generated. Instead, the electrical current's energy is converted directly into light, which is why LED lightbulbs are much more energy efficient.*

To turn an LED into a semiconductor laser, the device is structured to delay the spontaneous recombination of the electrons and holes in the p–n junction region. Then, one photon of light with exactly the right energy stimulates the recombination of an electron and hole, inducing the creation of a second photon of light at the same moment. These two photons will thus be exactly in phase (their electric and magnetic fields will oscillate in unison), and they will furthermore induce light

---

* All light from an LED will be of the same energy (and hence wavelength) determined by the difference between the bottom of the balcony and the top of the orchestra. It takes several different LEDs to make a white-light LED, each emitting light of a different color, which appears white when combined.

emission for the next electron and hole that enter the p–n junction region. Some of the light travels parallel to the junction, without passing through either the n-type or p-type semiconductor layers. The spontaneous emission of light is slow in this region. To speed up the stimulated emission, edges of the junction are polished like a mirror so that this light is reflected back and forth, continually stimulating the incoming electrons and holes to recombine and emit light in phase with the re-flected light beam. A small aperture with a focusing lens on one of the polished sides lets some of the light out, all of a single wavelength and in phase. It's all traveling in the same direction,* and this beam of coherent light is termed a "laser," an acronym for *light amplified by the stimulated emission of radiation.*

The laser pointer has a small battery to provide the voltage that forces the electrons and holes in opposition to the built-in p–n junction electric field, thus generat-ing the laser beam. The first semiconductor laser diodes emitted light in the infrared portion of the spectrum (not too useful for a pointer during a presentation, but good enough for CD players). Advances in materials research

---

* A laser beam exhibits very little spreading from the source of emission, such that it appears invisible from the pointer until it reaches the screen. Ir-regularities in the screen scatter the light in all directions, so that we can see it. To view a laser beam in transit from the pointer to the screen, you must first create small particulates to scatter the beam into your field of vision. In the classroom, one could beat two chalk erasers together, creating a cloud of dust, or use the smoke from a lit cigarette—but both chalk erasers and ciga-rettes have been banished from classrooms and conference rooms.

and manufacturing led to the development of visible red-light semiconductor lasers (used in DVD players), then green, and finally blue-violet semiconductor lasers (Blu-ray players) that are reliable and (most important) inexpensive. The laser scanners used to read bar codes run a laser beam across the black and white stripes, and measure the reflected light from the white regions. This generates a sequence of high and low light intensities that encode information, just like the arrangement of magnetic regions in the stripe on your credit card. The QR symbol on your smartphone does the same thing, but it can represent a hundred times more information, being a two-dimensional sequence of white and dark regions, rather than the one-dimensional bar code.

*You begin speaking, thanking everyone for taking the time to be here. Your host points out that it is difficult for those in the back to hear you. A small clip-on **microphone** is quickly procured, and the speaker system is turned on. You attach the microphone to your shirt and, after verifying that the sound level is not too loud, you again begin your presentation.*

A microphone is a device for converting sound waves into electrical voltages in such a way that the information encoded in the sound's changing amplitude and wavelength is preserved in the voltage variations. These voltages can then be electronically amplified, and when sent to a speaker, the original sounds can be made much louder. The development of techniques for translating variations of air pressure into voltages led to changes not just in the technology of telephones, broadcasting, and recording, but even in styles of popular music.

Microphones in early models of landline telephones would exploit the electrical properties of soot, and consisted of a cylinder containing small grains of amorphous carbon particulates, randomly packed together. These grains were jammed into the cylinder, with the top lid connected to a fabric membrane. When you spoke into the receiver of the phone, the variations in air pressure from your voice would cause the membrane to oscillate. Higher pitches would cause the membrane to vibrate back and forth faster than lower ones, and the amplitude of vibrations would reflect how loud you were speaking. Downward strokes would compress the carbon grains, while the pressure on the grains was released on the upswing. The contacts between grains would change depending on how hard the cylinder was squeezed. Carbon grains are pretty good conductors of electricity (not as good as a metal, but better than an insulator), and the ability of the cylinder filled with carbon particles to conduct a current when a voltage is applied across the cylinder is very, very sensitive to the average spacing of the grains. Two grains that are not in contact will not let a current flow, and those that are pressed tightly together will have a lower resistance than those that barely touch. When you speak into the microphone, the resistance of the collection of carbon grains varies precisely in proportion to the changes in air pressure from your speech, and the resulting modulations in the current passing through the cylinder will, when converted to a voltage, provide an electrical representation of your sound waves.

However, while it is adequate for speech, the frequency fidelity of the carbon-grain microphone is somewhat lacking. Any settling of the carbon grains would produce random changes in the current, which is what creates the crackling noise associated with these types of microphones. Research and development in the 1920s and 1930s of an alternative microphone based on capacitors gave a significant boost to a new type of popular singer. Unlike the older generation of songsmiths, for whom *projection* was a key to success (before the carbon-based microphone, singers would use a megaphone to ensure that their voice was heard at the back of the auditorium), singers were now able, thanks to the capacitor-based technology, to move closer to the microphone, crooning in a more intimate voice.

In a condenser microphone ("condenser" is just another term for "capacitor"), the membrane that vibrates due to incident sound waves is one plate of a parallel-plate capacitor. The ability of this device to store charge is highly sensitive to the separation of the plates. When this membrane flexes back and forth due to changes in air pressure, it decreases or increases the distance between the two plates. For a given amount of charge stored on the capacitor, this leads to a small but measurable change in the voltage across the capacitor. Capacitors can charge and discharge very quickly, so a properly designed circuit can quickly and accurately follow the variations in the plate spacing over a very wide range of frequencies. These microphones are very sensitive, and they can accurately detect very low-amplitude sound waves, as very little air mass needs to move to induce a measurable change in the voltage across the capacitor. Singers no longer needed to shout to be heard, and popular-music titans like Bing Crosby and Frank Sinatra had advancements in applied physics to thank for their careers.

Early radio stations would also use "ribbon microphones," which employed a metal sheet moving in the magnetic field of a permanent magnet. The physics is essentially the same as when a conventional speaker is run in reverse, only instead of a coil of wire, the microphone uses a thin metal strip. Changes in the amount of magnetic field passing through the area of the metal strip induce a current along its edges. (The metal strip is acting like a coil of wire, where the change in magnetic flux through the coil induces a current in the wire.) As the magnetic field threading the metal strip varies in time and amplitude from the incoming sound waves, these variations are reflected in changes in the current in the strip. Any possible note of the human voice could be accurately captured thanks to the wide frequency range of the ribbon microphone.

A microphone that requires less power to operate is desirable when worn on a lapel, or in mobile phones and other electronic devices where the energy source is a battery with limited capacity. The "electret microphone" ("electret" is a portmanteau combining *electricity* and *magnet*) is essentially a condenser or capacitor microphone, except that one of the capacitor plates is the electrical analog of a magnet, which carries a permanent electrical charge. A permanent magnet, such as iron, is made up of a multitude of smaller magnets from the individual iron atoms. Each atom has both a north and a south magnetic pole. When most of the atoms' magnetic fields point in the same direction, the solid piece of iron will have a large net magnetic field. An electret is similar, except that it has electrical dipoles (that is, a positive and negative charge separated by a small distance) instead of magnetic dipoles. Quartz (the crystal configuration of silicon dioxide) is an electret, as are many long-chain carbon-based polymer molecules.

There are three major advantages to incorporating electret materials in a condenser microphone: (1) they are cheap; (2) they are always polarized, so that no external power supply is needed to maintain a voltage across the capacitor, where one plate is the electret; and (3) they are cheap! If an electret is used as one of the plates of a capacitor, then, as it already carries an electrical charge, no external charging is necessary. The first and last redundant points are obviously very important when dealing with mass-produced consumer electronics. Although they do not have anywhere near the frequency fidelity of larger-scale condenser microphones or ribbon microphones, electret microphones can be small, lightweight, and inexpensive. Consequently, nearly every portable electronic device you own that can function as a telephone, including your smartphone as well as noise-canceling headphones, most likely employs an electret microphone.

*You begin your presentation: "Thank you all for coming. As you know, everyone loves handheld personal electronic devices, but some folks are concerned as to whether these products are safe. For example, people want improved phone reception, but they are nervous about being near cell phone towers. My goal today is to provide background information about what **radiation** is, so that we can assist our clients in making informed decisions about this topic."*

Any energy emission—whether of sound, electromagnetic waves, or high-speed subatomic particles (electrons, protons, neutrons, or combinations thereof) ejected from a nucleus—is termed "radiation" by physicists. The music you listened to from your speakers this morning radiated to all points in the room, so that you could hear it as you moved around while preparing breakfast. The information requested by your smart tablet as you checked your flight's status was broadcast from a wi-fi router radiating electromagnetic waves in the radio portion of the spectrum. Images of your ankle were taken using x-ray radiation. When passing through TSA, your luggage received additional screening via the explosive-trace detector, which used nuclear radiation to electrically charge the molecules to be measured in the ion mobility spectrometer. Radiation can be harmful, it can be useful, or it may not do anything at all. The key issue isn't the radiation, but what it does when incident on molecules. If it removes one or more electrons—so that the resulting molecule is now an electrically charged positive ion—then this is termed "ionizing radiation." As a general rule, any radiation of energy equal to or less than that of visible light is non-ionizing.

The atoms in any molecule are held together by chemical bonds, and the energy needed to break a chemical bond is roughly a few electron-Volts.* Some chemical bonds can be snapped with much less energy, and some are much tougher and require many electron-Volts to break, but if you said it takes a couple of electron-

---

* An "electron-Volt" is a unit of energy used by physicists; one electron-Volt is the gain in kinetic energy when one electron is accelerated by a voltage difference of one Volt.

Volts to break a bond, usually you would not be far off. A chemical bond involves interactions between the electrons from two neighboring atoms, and if incident radiation has enough energy to remove an electron, then the bond is compromised. A radio wave photon has an energy of a billionth of an electron-Volt, a microwave's energy is a millionth of an electron-Volt, and the energy of an infrared light photon is between a hundredth and a tenth of an electron-Volt. Any electromagnetic radiation in this part of the electromagnetic spectrum is non-ionizing. Visible light photons have an energy of two to four electron-Volts, depending on the color; an ultraviolet photon's energy amounts to approximately ten to a hundred electron-Volts; x-rays have an energy of a thousand electron-Volts; and the highest-energy region of the spectrum, termed "gamma ray" light, has energies of about a million electron-Volts. X-rays, gamma rays, and some ultraviolet light rays fall into the category of ionizing radiation. Then there is the energy emitted in the form of high-speed particles from a nucleus, which typically have energies of five to ten million electron-Volts, and is most definitely ionizing radiation.

"Wait a minute," someone in the meeting room interrupts. "Microwave ovens just use microwaves to cook food quickly. Then why do we say, when using one, that we are 'nuking' our food, if it has nothing to do with nuclear radiation?" You nod,

note that this is an excellent question, and reply: "Microwave ovens were developed by scientists and engineers who use the term 'radiation' for any type of energy, from fast-moving particles emitted from nuclei, to the glow of light from a candle or an incandescent lightbulb. Microwave ovens use electromagnetic waves created by a device that is electronically similar to a radio broadcaster. The microwaves generated in the oven induce polar molecules—such as the water in your food—to move rapidly back and forth, through interactions with the microwave's electric field, just like a compass needle can be made to point in a given direction by an external magnetic field. The microwave's electric field oscillates with a frequency of several billion cycles per second, and the water molecules flip at this rate, transferring this rotational kinetic energy to other molecules throughout your food. Microwaves and radio waves easily pass through solid objects (which is why we can pick up a radio station or get a cell phone signal when we're inside a building), so the microwaves penetrate into the interior of the food and all points heat at the same time—rather than the heat having to start at the outside and diffuse into the center—which is why the cooking time is much shorter than in standard ovens. Some expressions, such as 'nuking your food,' just caught on, but the microwave in your kitchen has nothing to do with nuclear physics."

To understand why some nuclei emit radiation, we first have to address why they stay together in the first place. In the heart of all atoms is the nucleus, consisting of a number of protons and neutrons packed into a very small volume. All the protons have a positive charge, while the neutrons are electrically uncharged. The positively charged protons repel each other, and they should fly away from the nucleus as soon as it is formed. Fortunately there is a stronger force operating within the nucleus, holding it all together. This force is called the Strong Force, and while it is approximately one hundred times stronger than the electrical force pushing the protons apart, it operates only over very short distances (roughly the diameter of a neutron). This is fortuitous, because if it had a long range, then such an intense attractive force would pull the nuclei from *all* atoms together into one giant sphere of protons and neutrons. So, if the Strong Force were too weak, all nuclei would fly apart due to the repulsion of protons, and if the Strong Force were too strong, all nuclei would be pulled into a single mass.

The larger the nucleus (that is, the more protons and neutrons it contains), the harder it is to pack them all together, balancing the repulsion of the protons with the very short-range attraction of protons and neutrons via the Strong Force. In fact, many nuclei do indeed fall apart, either partially or completely, or at least settle down in a more stable configuration,* with the emission

---

* In physics, when we say that a given configuration of protons and neutrons in a nucleus is "stable," what we really mean is that it has a lower energy than its starting configuration. A boulder can be balanced on a hilltop indefinitely,

of radiation as a consequence. These nuclei are termed "radioactive."

Once a nucleus has moved to a lower-energy, stable configuration, it typically is no longer radioactive. However, a cubic inch of uranium contains nearly a trillion trillion atoms, and at any given moment some of them will be emitting radiation. The precise moment when a particular unstable isotope will emit radiation is random, and can happen right away, or much later. It's not unlike rolling dice—one can get a "seven" in the first try, or only after many rolls. The "half-life" is the time one has to wait until, on average, half of all the nuclei in a material have emitted their radiation. In general, the quicker the nuclei decay, the shorter the half-life and the sooner the mass will exhaust all its radioactive isotopes and no longer be able to emit radiation.

The energy needed to confine protons in close proximity inside the nucleus is very high, so when an unstable nucleus relaxes to a lower energy state, the energy it gives off in the form of radiation is also very large. The predominant forms of radiation emitted from unstable nuclei are in the form of two protons and two neutrons tightly bound (called "alpha particles"), high-speed electrons (called "beta particles"), or gamma or

---

but the more stable configuration is when the boulder is down at the base of the hill. If the nucleus goes from one configuration with a higher energy to a more stable state with a lower energy, it must eject the energy difference somehow, in the form of radiation.

x-ray electromagnetic waves. This nomenclature dates from their discoveries at the turn of the twentieth century, before the inner workings of the atom were well understood.

It is very easy for either the positively charged alpha particles, the negatively charged beta particles, or the gamma rays, when they strike some other matter, to lose a small fraction of their kinetic energy and eject some of the electrons from the target atom, continuing on at high speed, ready to ionize the next atom they encounter. Some explosive-trace detectors employed by the TSA use a form of nickel that emits beta particles, while other versions use a heavier atom, americium, to emit alpha particles and ionize the molecules to be tested in the ion mobility spectrometer.

While ions occur naturally in the cell, too high a concentration can cause genetic damage, which is why high exposure to ionizing radiation can lead to severe illness and death. Of course, if the cells are already dead, as in meat or vegetables to be sold at market, then exposure to radiation will not cause further harm—but it *can* kill any still-living bacteria on the food. One way to eliminate the bacteria responsible for salmonella and other diseases is to expose the food to ionizing radiation (gammas, x-rays, or beta particles). This accomplishes the same result, as far as killing unwanted bacteria, as does cooking, without having to raise the temperature of the food. Food that has been exposed to radiation is described as being "irradiated." The damage done to the cells by the ionizing radiation does not result in new, unstable radioactive nuclei, and thus *irradiated food is not itself radioactive.* This confusion between "radioactive" and "irradiated" is unfortunate, with the very terms so emotionally charged that many will avoid any process that involves nuclear decay, even at the cost of preventable sickness and death.

CHAPTER SEVEN

You Go to a Hotel

*After your talk, you meet with several people in the office, answering some detailed questions. A small group takes you out to dinner at a nice local restaurant, and the wine with dinner is particularly good. After dessert they walk you to your hotel, conveniently located a few blocks away. During your presentation you were so focused on your talk that you had forgotten about your ankle, but now the painful twinge has returned. You are ready to get to your room and crash after your busy day. You provide a credit card and driver's license to the clerk at the front desk, who then prepares the registration form for you to sign. You glance at your **fitness monitor wristband** and see that, despite spending a good* *portion of the day on planes, trains, and automobiles, you have exceeded your daily steps goal. After you sign the registration form, the clerk gives you your room key: a blank card the same size as your credit card, in a small cardboard sleeve with your room number written on it. You head over to the bank of elevators. After a short wait the elevator arrives and swiftly takes you to your floor.*

When you take a step, you move in one or two dimensions (when traveling in a straight line or turning a corner, respectively), or in three dimensions (when climbing or descending a spiral staircase, for example). Every step involves a start and stop of motion, and any change in velocity is characterized by an "acceleration" (defined as the rate at which the velocity changes, whether increasing or decreasing). A fitness monitor that tracks your steps measures minute changes in your acceleration in three dimensions, and when those changes cross certain preset thresholds, it records that a step has been taken, adding it to a running tally.

Even when walking at a steady, uniform pace, you are constantly accelerating, as your foot presses against the ground, exerting a force that is reciprocated by the ground, pushing back on you. From an initially stationary position, your foot lifts up off the ground, and also moves in the direction you wish to travel. At the step's conclusion, your foot returns to the ground and, again at zero velocity, you are ready to take another step. Newton's second law of motion is expressed mathematically as: *Force is equal to the mass of an object multiplied by the acceleration.* Thus any change in motion, that is, any acceleration, whether an increase or decrease in speed or even just a change in direction, requires the action of a force. Your fitness monitor measures these forces associated with your steps using accelerometers that exploit the properties of harmonic oscillators.

At the start of this day we discussed alarm clocks and digital timers in terms of an oscillating mass on a spring, and a pendulum bob swinging back and forth. These oscillators had a natural frequency that was determined by their properties (how heavy the mass, how stiff the spring, the length of the pendulum string, and so on). But the sudden, brief application of a force can abruptly change this frequency of oscillation, as anyone who has pushed someone on a playground swing can attest. The fitness wristbands replace a swinging pendulum or a mass on the end of a coiled spring with a tiny cantilever. This cantilever, shaped like a miniature diving board, is about 500 microns long and 50 microns thick—where an average human hair has a diameter of approximately 100 microns—and any sudden movements are reflected in its deflections.

As the cantilever moves up and down in response to an external force, its mechanical deflections are measured and converted into an electrical signal through capacitor plates above and below it. The topside and underside of the tiny diving board have a metal plate, and above and beneath the board are other rigid metal plates, so that the board is half of two separate capacitors. When the board moves downward, the spacing between the bottom plates decreases, while at the same time the gap between the top plates increases. For a fixed amount of charge on the plates, the voltage across the capacitor is very sensitive to the plate separation, and changes in the board's position of no more than the size of a single blood cell can be detected. Insulating bumpers above and beneath the board prevent it from crashing into the plates during a large jolt.

Three such accelerometers oriented at right angles to one another monitor your three-dimensional motion. Advances in micro-machining and semiconductor processing have enabled the fabrication of accelerometers on a single silicon chip, which can be incorporated directing into micro-circuits.* Alternatively, the displacement may be measured by having the end of the board (or an actual mass on a spring in larger devices) be in direct contact with the detector. In this case the greater the force, the more it presses on the detector, which here could be a piezoelectric material or a piezoresistive crys-

---

* Such devices are called MEMS—for "micro-electrical mechanical systems."

tal (a material that changes its electrical resistance when strained).

The output of the accelerometer is a voltage that varies in time, responding to the sensor's motion. How this voltage is converted into a number of steps involves processing of the signal via computer algorithms that are, for the most part, proprietary information of the various companies that market these monitors. Using these protocols, the monitor will calculate the number of steps, how many calories are burned, and your pulse rate—though this last requires a separate measurement using light-emitting diodes.

To record your pulse, your wrist monitor shines light from an LED onto your wrist and uses a photodetector (a semiconductor whose resistance is sensitive to the amount of illumination) to determine how much light is reflected. Different wavelengths of light are used, which penetrate to different depths beneath the skin. The light scattered back to the detector will vary with changes in the mass under the skin, such as occur with the periodic flow in your blood vessels. Correlating the voltages from the optical detector with those from the accelerometers helps correct for some (though not all) of the noise in this measurement. Anything that is not a blood vessel (such as bone, tendons, and wrist tattoos) will also scatter light and obscure the desired information. A much better location for such pulse monitors is your fingertip or earlobe, where the skin is thin and has few obstructions, which is why during your doctor's visit your pulse was measured with a clamp on your finger that employed the same physical principles as the wrist monitor. A lot of technology is combined to convert mechanical actions into digital electrical signals. It may be one small step for you, but it is one giant leap for science.

*You arrive at your floor, a little jostled by the sudden deceleration of the elevator car. Checking the sign on the wall, you head down the long hallway to your room. The corridor is initially dark, with the ceiling lights turned off, making it difficult to make out the room numbers. However, as you walk farther, a **motion sensor** detects your presence and switches on the lights.*

Most motion sensors are forms of infrared night-vision goggles, detecting the light emitted by your body and converting this electromagnetic radiation into an electrical signal. You emit radiation for the same reason your toaster wire glowed red when you prepared your bagel this morning: the combo platter of changing electric currents generating changing magnetic fields, and changing magnetic fields inducing electric currents. The motion sensor detects this infrared light using a detector that employs a physical phenomenon related to that which operates in the imaging wand during an ultrasound scan.

First let's consider the glowing source of infrared radiation—you. You are made of atoms that are in constant motion. I'm referring not to blood flow or any large-scale transportation, but simply to the oscillations of the atoms about their positions in the molecules in your cells.

Consider an organic molecule consisting of rings of carbon atoms with hydrogen, nitrogen, and oxygen atoms. Isolated from the rest of the universe, floating in outer space, it would eventually find itself in its lowest energy configuration, and aside from a slight tremor due to quantum mechanical effects,* it would be stationary. If the same molecule were placed in a pot of hot water, the high-kinetic-energy water molecules would continually bombard this organic molecule, transferring some of their kinetic energy to it. Eventually, the organic molecule would have the same average energy as the water molecules that surround it, and there would be no further net transfer of energy. But then the organic molecule is in a higher energy state, and this energy is spread out to all of its constituent atoms. Each atom now has

---

* Even at the absolute zero of temperature there is some jittering of the atoms, called "zero-point motion," that is a consequence of the Heisenberg uncertainty principle. This is an interesting phenomenon, but it has nothing to do with the operation of motion sensors, so forget I brought it up.

a share of kinetic energy but is constrained from leaving the molecule by the chemical bonds holding it in place. This kinetic energy manifests as the atoms vibrating around their positions in the molecule; and as they shake, so do their electrons.

The organic molecule in this scenario is a stand-in for all of the molecules in your body, with the pot of hot water being your body. The food you eat contains molecules with stored chemical potential energy that through a series of chemical reactions is converted into excess kinetic energy of other molecules, used to activate all the chemical and biological processes in your body. Roughly half of the energy you ingest in a typical day goes toward any large-scale activities (walking, running, making breakfast, passing through TSA at the airport, etc.), while the other half keeps you alive, even if you remain motionless in bed (keeping your heart beating, your lungs breathing, and maintaining your body temperature). Temperature reflects the average energy per atom of an object, and as you are reading this, the atoms in your body are all vibrating with an average kinetic energy described by a temperature of 98.6°F, and their vibrating electrons emit light in the infrared region of the electromagnetic spectrum. We are all, due to our natural body temperature, shining 100-Watt lightbulbs of infrared light.

The motion sensor in the hotel hallway detects the

infrared light you are sending using a semiconductor device called a "pyroelectric detector," which converts heat (*pyro*) into a voltage (*electric*). This effect is similar to piezoelectricity, which plays an important role in maintaining the timing of computer chips and in ultrasound scans. In the case of a piezoelectric device, a change in stress on the crystal creates an electrical polarization of the lattice structure that in turn generates a net electric field (equivalent to a voltage) across the material. In a pyroelectric material the absorption of heat causes the atoms in the detector's crystal structure to vibrate faster, and due to anharmonic effects of the chemical bonds and the unique arrangement of the atoms in the solid, a distortion of the lattice occurs, creating a net electrical polarization and hence a voltage.

The crystal does not have to be near an open flame for this to work. It will also generate a voltage if infrared light from a heat source (such as yourself) is close by. As most hallways and rooms are kept at a temperature of 70°F, your skin temperature is roughly 20 degrees warmer than your surroundings, and the energy of your infrared light is greater than the surrounding background.* But what if the air-conditioning goes off

---

* Night-vision goggles operate on the same principle—you are a bright source of infrared light compared to your environment, especially at night when the average temperature drops without sunlight exposure, and photoconducting semiconductors convert this infrared light into a voltage.

during the summer, and the hallway's temperature rises? Most sensors use two detectors that are wired in opposite ways. That is, a positive voltage difference that would generate a current in one direction, created by the absorption of heat from one detector, is connected to the negative voltage difference of the second detector, which would create a current in the opposite direction. Thus, as the hallway's temperature rises, as long as the additional heat is uniform, there will be no net signal sent out from the sensor. A person, on the other hand, is a very localized heat source, and most sensors are part of a circuit that sends out a voltage to turn on the lights only when there is a change in the profile of the heat signature detected, such as when you walk down the hallway.

Such motion sensors are termed "passive," as they detect the infrared light only from an external source—such as your body. They use less power this way. There are also "active" motion sensors, which send out microwaves and look for changes in the echoing signal that returns to the detector. More sophisticated motion sensors that are part of certain residential security systems employ both passive infrared light detection and active microwave broadcasting. Such systems can use simple computer chips that will not trigger if the moving object is below a certain mass threshold, avoiding false positives due to pets.

You stop at the door, but there is no slot for you to insert your room key card. That's because the clerk gave you a **proximity card**. *You touch the card to the circular pad underneath the door handle, but the small LED flashes red, indicating that you can't enter the room. You try again, and again get the red signal. A wave of frustration that you will have to go all the way back to the lobby and have your card reprogrammed quickly disappears when you double-check the room number and discover that you have accidentally stopped at the wrong door. Yours is two rooms farther down the hallway, and when you repeat this process there, you receive a green LED. You hear the bolt of the door's lock being thrown back and can now enter your room.*

The proximity card used as hotel room keys and as reloadable mass-transit fare cards is essentially an E-ZPass system, miniaturized to the size of a credit card.

We've encountered the basic principles of the passive card-entry system many times throughout the day, starting with the recharging of your electric toothbrush, in which an alternating current in one coil creates an oscillating magnetic field, which passes through a second wire loop, inducing an alternating current in the second coil. In the case of the hotel's proximity card, one coil is in the key card reader in the door-handle assembly, connected to a power supply. Have you ever noticed that such readers, on doors or train station turnstiles, are raised circles or large squares? The better to house a large coil that acts like a radio antenna. The second coil is inside the key card, running along the perimeter of the card. This coil, together with a capacitor inside the key card, acts as a radio receiver antenna that is tuned to only one station—the frequency being broadcast from the card reader in the door. The broadcaster in the door sends out a low-frequency radio signal that induces a current in the key card's loop. This current in turn powers the microcomputer chip also embedded within the card. This chip sends a digital code to the coil in the card, which in turn broadcasts a message to the coil in the door. This really *is* like the E-ZPass system,* though the receiver and detector have to be within centimeters of each other, rather than the relatively vast distance of many meters at a tollbooth. Upon receipt of this digital code, the chip in the door's key card reader sends an electrical signal to a "solenoid" (basically an electromagnet). This electromagnet generates a force that pulls the latch of the door open, allowing you to enter your hotel room.

Unlike the key card, the reader in the door can recognize multiple codes, which is convenient for housekeep-

---

* Some systems also share a feature with your remote entry system for your car, in that the digital code sent by the card is generated by an algorithm that updates the code when the card is used. This prevents a key card from the room's previous occupants from working once their reservation is over.

ing workers who need to gain access to all of the hotel's rooms but would prefer not to carry hundreds of individual key cards. A timer is set for the key card's digital code. Once your reservation is over (or your account is depleted, in the case of proximity cards used in mass transit systems),* the code on the reader is changed, and your key card will no longer grant you access.

These passive proximity cards must be brought very close to the card reader, which transmits the radio wave signal to the card's coil antenna. Active proximity cards can broadcast their serial numbers and digital codes over larger distances, and are used for automatic toll paying or opening security gates. The greater energy demands for these active cards are often provided by a thin lithium battery embedded within them, which is why these cards are usually thicker than a business card.

*You unpack your bag, which doesn't take too long. You take a quick shower, and are grateful for the fluffy bathrobes the hotel provides. Using your tablet, you access the hotel's wi-fi. You check your e-mail, then pick up the **remote control** device from the bedside table and turn on the television set mounted to the wall.*

---

* Many cities still encode information on their fare cards via a magnetic stripe, which must be swiped to be read, employing the same physics as a credit card reader.

To make a remote control device, one needs a light-emitting diode in the control unit that shines a beam of infrared light, and a detector in the TV set. An infrared LED is much easier and cheaper to manufacture than one that emits in the visible part of the spectrum. Similarly, most commonly used semiconductors strongly absorb infrared light, which is convenient. The LED is constructed so that it emits infrared light at one particular wavelength, and a filter at the television set will let only that wavelength pass to the detector. (This means you don't have to worry about the infrared light emitted by your body changing the channel.)

When you press a button on the remote control, you effectively close a switch in the printed circuit underneath the set of buttons. In a printed circuit, a thin sheet of an insulating board has various metallic lines printed on one side. The rubber button on the controller unit has a metal disc on its underside, and when you press the button the metal disc touches the printed circuit board and closes a connection, and a current is sent to the infrared LED. A set of on/off flashes of infrared light, in which information for the set is encoded, is then emitted.

This signal is detected by a semiconductor in the television set that has a very high resistance in the dark, but a very low resistance when it absorbs light. In the dark, the resistance is so high that practically no current can flow through the semiconductor; it is as if a switch is left open. When infrared light shines on the detector in the television set, electrons are promoted from the lower-energy filled band to the higher-energy empty band. Now a current can flow—like a switch has been closed, completing the circuit. The on and off flashes effectively open and close the switch in the television, and the on and off currents are the information for the computer chip in the set. The chip interprets the message encoded in the current, and responds by sending the requested signal to other parts of the television circuit, changing the channel or altering the volume.

A striking consequence of quantum mechanics is the band structure of solids—as in the analogy of a filled lower-energy orchestra and empty higher-energy balcony. Different elements or molecules will have different energy spacings. Quantum mechanics indicates that an atom's electrons can have only very specific values of energy, with a gap in energy between the last allowed energy state that has electrons and the first possible energy value that is unoccupied. When many atoms are brought together, these energy states broaden into bands. In semiconductors and insulators the lower-energy band is full and the first upper-energy band is empty, while in metals the lower-energy band is only partially filled (which is why the electrons in metals can carry current even without moving into the balcony states). Any light that does not have enough energy to bridge the gap between the orchestra and the balcony cannot be absorbed, no matter how bright, and passes through the material as if it were not there, a direct manifestation of the quantum nature of matter.

Infrared light is associated with heat, because it is easily absorbed and converted to kinetic energy by most molecules. In fact, most of what we know about molecules is thanks to the quantum mechanical nature of how they absorb infrared light. The quantum partition of allowed energy states into occupied and unoccupied bands in solids exists within molecules as well. Think of a very simple molecule, such as two hydrogen atoms bonded together to form a hydrogen molecule (H_2). The H_2 molecule can move from one place to another, and quantum mechanics places no constraints on its energy because it is in free space. But there are other ways it can move. It can vibrate back and forth along a line connecting the two hydrogen atoms, and the energy of this vibration can take on only specific values. In addition, it can rotate about an axis passing through the midpoint between the hydrogen atoms, and here again, only certain energies of rotation are possible. These energy spacings for rotations and vibrations are usually in the infrared (or lower energy) portion of the spectrum, and the molecule will absorb infrared light only if it has exactly the right energy to induce one of these transitions. Different molecules will have different rotation-vibration energy spacings, and in fact each molecule has a unique "infrared spectrum fingerprint." One of the most powerful tools physical chemists and forensic scientists have for determining the chemical composition of an unknown object is to measure its infrared spectrum and compare it with that of known molecules.

The silicon dioxide molecules in window glass can also vibrate around their locations in the solid, and these vibrations lead to absorption of infrared light. Ultraviolet light, with energy greater than the violet end of the visible light spectrum, will induce transitions between the orchestra band and the balcony band; the glass will absorb it. So, we can look through glass because it transmits visible light, and in fact, visible light is the only light that glass will pass.*

The set comes to life, and as you change channels, searching for the local news, you marvel at the bright, high-contrast image. You are even more impressed when you take a closer

---

* Note that carbon sits right above silicon in the Periodic Table, and carbon dioxide is chemically similar to silicon dioxide, and will also let visible light pass unmolested but absorb infrared light. This is why carbon dioxide is termed a "greenhouse gas." Visible light during the day strikes the surface of the Earth, warming it. At night, the planet cools down by emitting infrared light. But if the light is absorbed by carbon dioxide (and other gases) in the atmosphere, some of the energy can then be re-radiated back to the Earth's surface, keeping it warm even when the sun is not out.

*look at the **flat panel television** set and notice how remarkably thin it is. Whereas the LCD television set you have at home is a few inches thick, this set seems to be no more than an inch from front to back. You notice an embossed label on the front of the casing that lists this set as an OLED.*

The transition from bulky cathode-ray tube television sets to flat panel televisions was enabled, in part, by the replacement of vacuum tubes by solid-state transistors. The newer forms of flat panel televisions make a similar, radical change in technology, where the transistor, light source, and liquid crystal display (LCD) are replaced by an ultra-thin light-emitting diode composed of organic molecules. These "organic light-emitting diodes" (OLEDs) can be painted onto surfaces using silk-screening techniques; they use much less power and have faster display times than LCD-based televisions.

To make an LCD television, as in the LCD projector described earlier, we need a light source (either a cold-cathode fluorescent lamp or a semiconductor-based light-emitting diode), a large array of liquid crystal pixels, and a thin-film transistor behind each pixel to process the signal sent from the cable or antenna and determine whether an electric field should be applied across the glass slides or not. The OLED display does all of the above with a single light-emitting diode that can be ten times thinner than the LCD system.

As described before, in an ordinary light-emitting diode, different chemical impurities are added to semiconductors that either supply extra electrons to the balcony or remove electrons from the filled orchestra, in essence adding holes to the lower band. Then these two materials are combined on top of each other. An electrical current is pushed through the two semiconductor layers, electrons from one side and holes on the other, and where they meet at the interface between the two layers, the electrons in the balcony drop down into the holes in the orchestra, emitting a photon of light when they do. An OLED works in exactly the same way, except that one organic molecule is used for the side that carries the electrons and a different molecule is used for the one that transports the holes. These two different molecules are separated by another layer, of a third organic molecule, that serves as the region where the electrons and holes recombine, emitting light. By changing the chemical composition of this third molecule, one can vary the color of light that is emitted, and OLEDs can shine in red, green, or blue light—or, when these primary colors are combined, white light.

In an LCD television display, depending on the electric field across the glass plates, the liquid crystal will either block or transmit the illumination from the light source, which is always on. A color filter then shifts the light to red, green, or blue, and when they are combined a color image can be displayed. In an OLED-based television these elements (aside from the thin-film, amorphous silicon transistor)* are replaced by the organic-molecule light-emitting diode. No crystalline semiconductor LED as a light source is needed, as the OLED emits its own light. No color filters are needed, as different OLEDs can be used for the three basic colors.** No bulky glass slides and liquid crystal are needed to block the light—when a pixel should be dark, the current is turned off through the OLED. In this way the black pixels on an OLED display are true blacks, providing much sharper contrast in the resulting image.

The greater the current passing through the OLED, the more photons are emitted and the brighter the image. The OLEDs require much less energy to run than the light-emitting diode and liquid crystal display combination, making these types of display very attractive for smartphones, tablets, and wrist fitness monitors relying on battery power. One can turn the OLED on and off hundreds of times faster than one can twist and

---

* Amorphous semiconductors are used, as the array of transistors needs to cover a large area for big-screen flat panel televisions.

** For technical and financial reasons, some OLED displays combine red, green, and blue OLEDs to generate white light, which is then passed through a filter to create either a red, green, blue, or white pixel. There's a benefit in the manufacturing process to doing it this way, but one can just as easily use the red, green, and blue OLEDs directly.

untwist the liquid crystal pixel, so OLED displays avoid the motion blurring that sometimes plagues LCD sets. The OLED pixel is thinner than the diameter of a single red blood cell, while the comparable functionality for an LCD pixel requires a few millimeters of thickness.

So why don't all flat panel television sets use OLED displays? They're still much more expensive to manufacture than LCD sets. In addition, the molecules used in the blue light OLED tend to degrade over time, forcing cumbersome workarounds until better molecules can be found. Finally, the organic molecules in the OLEDs don't react well with water—admittedly more of a concern for smartphone displays than wall-mounted televisions. Still, a large advantage of using organic molecules is that you can spread them over vast areas using either ink-jet printing techniques or silk screening, as for a T-shirt. As no heavy and inflexible glass sheets are required to confine a liquid crystal layer, the OLED displays can be printed onto plastic sheets that can be rolled up and stored in your back pocket. Or one could print the display directly onto clothing or other nontraditional surfaces. As research in this and other areas progresses, what once was considered science fiction will become commonplace.

After watching the news for a while, you settle into bed and, changing the channel, find a comforting old movie. It's one of your favorites, the 1985 film Back to the Future. *You catch the end of the film, where Marty and his girlfriend, Jennifer, are surprised when the time-traveling DeLorean appears in a flash of light in front of them. Doc Brown exits the vehicle, wearing "future clothes." Arguing that Marty and Jennifer must accompany him back to 2015, he refuels his machine using organic and inorganic matter (that is, garbage) that he places inside the "Mr. Fusion" reactor strapped to the hood of his car. With Marty and Jennifer in the car, and Doc at the wheel, Marty points out that they need to back up, as they don't have enough road ahead of them to get up to a speed of 88 miles per hour (the speed at which the car's time-travel mechanism will be activated). Putting on "future sunglasses" (that no one in 2015 has ever worn), Doc replies, as the De-Lorean transforms into a* **flying car**: *"Roads? Where we're going we won't need roads."*

There are only two reasons why, well into the twenty-first century, we still don't have flying cars: physics and chemistry.*

The physics constraint comes from the principle of conservation of energy. A car is a heavy object, and

---

* I am considering here the flying cars that have vertical lift-off and landing capabilities, and not ones with retractable wings that are essentially small planes. Versions of this latter "flying car" do indeed exist, but they are not the vehicles that have long been promised by science fiction films and comic books.

to lift it even a few feet off the ground means raising its potential energy by 15,000 Joules. For comparison, a 100-mile-per-hour fastball has a kinetic energy of 140 Joules. To lift the vehicle so that one truly is independent of any preexisting roads or highways would easily call for over one hundred times more energy. And this is just to lift the car and park it on a cliff.

To keep it elevated while driving, you have to continually exert a downward force, so that, thanks to Newton's third law of motion (forces come in pairs), the car experiences an upward force, counteracting the downward pull of gravity. For a DeLorean, this would require a constant downward force of about 2,800 pounds, exerted all the time. As soon as you stop providing this downward force, gravity will accelerate you back to Earth. We've discussed how one would generate this much force before—using an internal combustion jet engine. A single jet engine can easily provide sufficient thrust to support an automobile, and most imagined flying cars have four such engines, where the tires would normally be located. These levitating vehicles would presumably have much smaller jet engines than their larger cousins employed on modern aircraft, but let us assume that the reduced thrust of each single engine is compensated by the fact that there are four of them on our flying car. Whenever these sorts of flying cars appear in Hollywood movies, they are typically portrayed as quieter than conventional internal-combustion-engine automobiles—but in reality, if you had jet engines sitting at the four corners of your vehicle, the noise would serve as a serious distraction to your flying and tax the capabilities of the best noise-canceling headphones. Nor would it be too pleasant for any pedestrians standing underneath the exhaust of these jet engines.

But there is another limitation, derived from chemistry, that makes flying cars impractical. As just mentioned, it takes a considerable amount of energy to keep a car levitated in the air, and the vehicle needs to carry the source of this energy with it.* Here, as for conventional automobiles, the important issue is *energy density*—how much stored chemical potential energy can you carry per pound of fuel?

In *Back to the Future* the flying car converts the organic waste into energy through a "Mr. Fusion" device. Organic matter, such as banana peels and beer, is mostly water, and a certain number of the hydrogen atoms in water molecules contain an extra neutron in their nucleus, bound through the Strong Force, which is termed "deuterium." H_2O molecules with deuterium replacing the hydrogen are called "heavy water." If two deuterium nuclei are fused into a single larger nucleus of two protons and two neutrons, we would then have a

---

* In the *Back to the Future* films, this energy constraint is acknowledged, and in fact much of the plot revolves around obtaining energy (in the form of plutonium or lightning bolts) to operate the time machine.

helium nucleus. The helium nucleus has a mass slightly less than two deuterium nuclei, and the change in mass is converted into a large amount of energy (through $E = mc^2$) in this fusion process. This nuclear fusion occurs in the heart of the sun and is the source of all of its energy (re-created on Earth in the form of the hydrogen bomb). If we could achieve the same effect in a small device that could fit on a kitchen counter, we truly would be in the World of Tomorrow, and flying cars would be the least of the changes to our lives.

It's not just how much energy your flying car can carry in its fuel supply, but how fast the energy can be deployed. Power is not the same as energy, but rather is a measure of the speed or rate of energy conversion or use.

Being able to rapidly provide energy is important if you are going to have a flying car. Unlike airplanes that soar at altitudes of over 30,000 feet, where the thin air reduces the effect of air drag, a flying car will need to expend considerable energy pushing the air out of its way, in addition to providing a vertical force to counterbalance the downward pull of gravity, and have some left over to actually fly the vehicle to its destination. The ideal power supply for a flying car would have a high energy-per-pound combined with the ease of storage of a lightweight lithium battery. Recent research on a device that has appeared time and again throughout your day (and in this book) may offer a breakthrough on energy storage and power delivery. It is a simple construct that is the foundation of much of the technology you encounter on a daily basis: the capacitor.

In a lithium battery the ions that provide the electrical energy are generated via chemical reactions, involving the motion of atoms. As small as a lithium ion is, it weighs 12,000 times more than a single electron. Instead of having the stored electrical charges be associated with ions, it would be much more efficient if we got rid of the heavy nuclei and just used lightweight electrons to store electrical energy. This is exactly what a capacitor does, with the added benefit that when you draw a current from the charges stored on a capacitor, the electrical signal moves at nearly the speed of light, much faster than the rate of the chemical reactions that charge up the lithium battery's electrodes. So the power (that is, the rate of energy provided) delivered by a capacitor is very high. Where capacitors fall short is in the total amount of energy they can store. Up until recently, the best energy densities (Joules per kilogram) one could achieve in a capacitor were over 1,000 times lower than in the best lithium-ion batteries. In other words, even though they could discharge or recharge much faster than a battery, they were not viable as power supplies for flying cars (or even regular ones, for that matter).

Newer capacitors under development may change all that, dramatically increasing the amount of electrical energy they can hold. In these capacitors, the plates

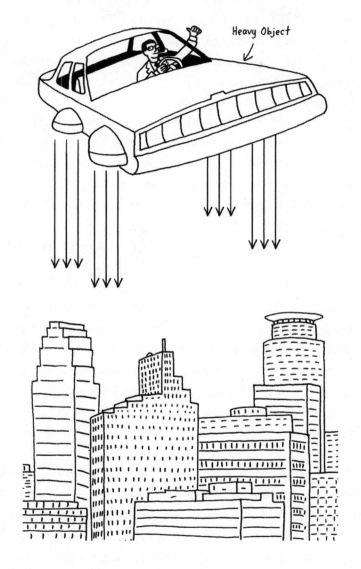

FIGURE 7

are coated with a very thin (about a tenth of a millimeter) layer of carbon. Using chemical tricks, this carbon is roughened up so that instead of a smooth, uniform surface it looks like the interior of an English muffin, with many nooks and crannies. Such carbon is termed "activated"; any process that proceeds on the material's surface, such as storing a charge, will be dramatically improved by having many more places on the surface with which to react. The effective surface area of capacitors with activated carbon is nearly 10,000 to 100,000 times larger than that of the smooth metal plate, and the stored charge density correspondingly increases. In addition, the space between the activated carbon-coated metal plates is filled with a fluid, called an "electrolyte," that contains both positively and negatively charged ions. By piling up ionic charges right next to the plates, the electrolyte reduces the effective separation of the charged plates to roughly the length of an atom, greatly increasing the electric field between the ion layer and the electron layer. Such devices can theoretically store over ten million times more electrical energy than a comparably sized normal capacitor, and are termed "ultra-capacitors," or "super-capacitors."

While the basic physics of super-capacitors has been known since the 1960s and '70s, advances in materials science in the 1990s have brought their power densities close to the levels required for use in cars (the Earthbound versions). Present super-capacitors have an

energy density roughly 5 percent of a lithium-ion battery's and can deliver that energy in under a second, for
a power density, or rate of energy delivered per mass,
nearly ten times greater than for a lithium-ion battery.
A super-capacitor could recharge in seconds (limited by
ion diffusion over very short distances), compared to an
hour for a conventional lithium battery. Because there
are no chemical reactions to take place in the capacitor,
there is no degradation of the terminals, and the lifetime
of such a charge-storage device would exceed the life of
the car or truck that it powered. While they aren't ready
to serve as the main source of power for transportation,
super-capacitors are already being used in specific situations where a large burst of power is called for—in some
buses and trucks needing an acceleration boost when
climbing a hill, or for the rapid opening of emergency
exit doors on jetliners. And while it is not very likely,
perhaps one day enough energy could be stored in these
devices that we will indeed achieve freedom from the
tyranny of roads.

Throughout your day, you've relied on physics—
very old and very new. The principle of conservation of energy governs all aspects of our lives. Nearly
as important have been the basics of electromagnetism,
where electrical currents create magnetic fields, and
changing magnetic fields induce electrical currents. The

study of the properties of matter and light on the atomic scale led to the development of quantum mechanics, which in turn enabled powerful applications of chemistry and solid-state physics. These same basic concepts have been employed again and again throughout the day, helping make it anything but ordinary.

You turn off the OLED flat panel television using the remote control. Tapping the touch screen on the LCD of your smartphone, you check that the alarm clock is set to wake you tomorrow morning so that you have enough time to catch your return flight home. For the alarm you select a tune stored in the phone's computer memory, a snippet from a song you downloaded from the Cloud. Still thinking about the end of the movie, you drift off to sleep, grateful for the marvels of science, wondering if you'll ever get your flying car.

ACKNOWLEDGMENTS

The idea of a book explaining the basic physics underlying the technology we employ on a daily basis began with
a suggestion from Roger Scholl, my first editor at Crown.
The talented Domenica Alioto then piloted the resulting
unwieldly manuscript, rightly calling for serious prun-
ing. It was Domenica who proposed recasting the book to
follow someone through a hypothetical day; she coined
the term "narrative physics" to describe the book's structure. When Domenica was not available, Claire Potter
seamlessly took over the editing reins and helped bring
the book to completion. I am deeply grateful for all three
editors, along with the thoughtful contributions of the

copy editor, Lawrence Krauser, though I confess to being concerned as to what it implies about my writing that it would require the attention of so many talented and dedicated personnel. I also must thank my agent, Jay Mandel, who initiated the entire venture and has been tremendously supportive throughout this long process.

One of the pleasures of writing this book has been to observe myself as I carried out various tasks the eponymous "you" would perform, seeing the technology that was in plain sight that I nevertheless rarely noticed. I then had an opportunity to explore how these devices functioned, deepening my appreciation. Often my first step when researching many of the devices discussed here was to take down from my bookshelf Professor Louis Bloomfield's excellent *How Everything Works* (John Wiley & Sons, 2008), closely followed by visiting the websites www.howstuffworks.com and www.explainthatstuff .com. I also greatly benefited from discussions with my colleagues, particularly John Broadhurst, Paul Crowell, Michel Janssen, and Chris Kim. Brian Skinner and C. C. Huang provided guidance on the text for several of the devices, and E. Dan Dahlberg, Aaron Wynveen, Doug Kohrs, and Carolyn Kohrs graciously read the entire manuscript in various stages of its existence. Their input greatly improved the final text, though of course any errors or confusing points remain my responsibility.

And I owe a debt that cannot ever be paid to my wife and family. My wife, Therese, has been encouraging, supportive, and patient. The book simply could not have been written without her. Her inspiration and feedback throughout the writing process were essential, and I am thankful every day for her love. Honey, let's take that walk now.

NOTES

CHAPTER ONE: YOU BEGIN YOUR DAY

2   **A pendulum is a very simple device:** Paul A. Tipler and Gene Mosca, *Physics for Scientists and Engineers*, 6th ed. (W. H. Freeman, 2007), 465–74.

4   **Your electric company rotates coils of wire:** Roger A. Hinrichs and Merlin Kleinbach, *Energy: Its Use and the Environment*, 3rd ed. (Brooks/Cole, 2002), 358–67.

7   **specially designed chips that shift the frequency:** Paul Horowitz and Winfield Hill, *The Art of Electronics*, 2nd ed. (Cambridge University Press, 1989), 885–86.

9   **keeping time in a device:** Dominique Flechon, *The Mastery of Time: A History of Timekeeping, from the Sundial*

to the Wristwatch: Discoveries, Inventions, and Advances in Master Watchmaking (Flammarion, 2012).

9   **predates the existence of electrical power:** Streets were first electrified in 1882, while the Seth Thomas Cook Company patented a small mechanical alarm clock in 1876. Maggie Koerth-Baker, *Before the Lights Go Out: Conquering the Energy Crisis Before It Conquers Us* (John Wiley & Sons, 2012), 10.

11   **material is called piezoelectric:** Bernard Jaffe, *Piezoelectric Ceramics* (Academic Press, 1971); Louis Bloomfield, *How Everything Works*, 5th ed. (John Wiley & Sons, 2008), 300.

13   **the "musical conductor":** Alvis J. Evans, *Basic Digital Electronics* (Master Publishing, 1996), 24–26.

13   **switch in less than a nanosecond:** Laptop and desktop computers have CPUs that switch much faster than those in smartphones, but they also require more energy. One of the imperatives of smartphone engineering is to use as little energy as possible, to prolong battery life.

15   **convert electrical voltages into mechanical vibrations:** Bloomfield, *How Everything Works*, 426.

16   **The natural frequencies of wood:** Bruno L. Giordano and Stephen McAdams, "Sound Source Mechanics and Musical Timbre Perception: Evidence from Previous Studies," *Journal of the Acoustical Society of America* 121, 2384 (2007); D. Murray Campbell, "Evaluating Musical Instruments," *Physics Today* 67 (April 2014): 35–40.

19   **is called a "transformer":** Tipler and Mosca, *Physics for Scientists and Engineers*, 1004–1006.

22   **Toaster wire is usually composed:** Bloomfield, *How Everything Works*, 225.

24   **sugars and starches undergo a chemical reaction:** Joseph J. Provost, Keri L. Colabroy, Brenda S. Kelley, and Mark A. Wallert, *The Science of Cooking: Understanding the Biology and Chemistry Behind Food and Cooking* (John Wiley & Sons, 2016), 198–217.

25   **subverts the standard paradigm:** See, for example, Donald S. L. Cardwell, *From Watt to Clausius: The Rise of Thermodynamics in the Early Industrial Age* (Cornell University Press, 1971).

26   **a refrigerator is an engine run backward:** Mark W. Zemansky, *Heat and Thermodynamics*, 5th ed. (McGraw-Hill, 1968), 179–85. A description of evaporation cooling is presented in James Kakalios, *The Physics of Superheroes: Spectacular Second Edition* (Gotham Books, 2009), 170–71.

26   **Refrigerators once used Freon:** Bloomfield, *How Everything Works*, 261.

26n   **In 2016, 170 nations:** Coral Davenport, "Nations, Fighting Powerful Refrigerant That Warms Planet, Reach Landmark Deal," *New York Times*, October 15, 2016.

CHAPTER TWO: YOU DRIVE INTO THE CITY

30   **A hybrid car uses both:** Chuck Edmondson, *Fast Car Physics* (Johns Hopkins University Press, 2011), 194–200.

31   **A typical battery employs ions:** Ibid., 187–91.

33   **There are four steps in a combustion cycle:** Louis Bloomfield, *How Everything Works*, 5th ed. (John Wiley & Sons, 2008), 260–64.

34   **A gasoline molecule consists:** Gordon J. Aubrect, *Energy: Physical, Environmental and Social Impact,* 3rd ed. (Pearson Prentice Hall, 2006), 217–18.

35   **In some hybrids:** Edmondson, *Fast Car Physics,* 196–98.

36   **Your GPS device communicates with satellites:** See, for example, Patrap Misra and Per Enge, *Global Positioning System: Signals, Measurements and Performance,* rev. 2nd ed. (Ganga-Jumana, 2010).

38   **Albert Einstein's general theory of relativity:** There are many excellent popular science descriptions of general relativity, particularly Kip S. Thorne, *Black Holes and Time Warps: Einstein's Outrageous Legacy* (W. H. Norton & Co., 1995); and Pedro G. Ferreira, *The Perfect Theory: A Century of Geniuses and the Battle Over General Relativity* (Houghton Mifflin Harcourt, 2014).

39   **To paraphrase theoretical physicist John Archibald Wheeler:** Wheeler, a collaborator with Einstein, coined the terms "black hole" and "wormhole." In his book *Geons, Black Holes and Quantum Foam* (W. W. Norton & Co., 2000), 235, coauthored with Kenneth William Ford, he wrote: "Spacetime tells matter how to move; matter tells spacetime how to curve."

39   **There are *two* effects stemming:** Chad Orzel, *How to Teach Relativity to Your Dog* (Basic Books, 2012), 220–21.

41   **The E-ZPass system:** See "How E-ZPass Works" on HowStuffWorks website, http://auto.howstuffworks.com/e-zpass.htm.

41   **How does a radio work?:** Bloomfield, *How Everything Works,* 428–35.

45   **As the density of cars:** Books that describe models of traffic flow for a general readership include Mitchel Resnick, *Turtles, Termites and Traffic Jams: Explorations in Massively Parallel Microworlds* (MIT Press, 1994); and Tom Vanderbilt, *Traffic: Why We Drive the Way We Do (and What It Says About Us)* (Knopf, 2008). Those looking for more mathematical discussions should start with Dirk Helbing, "Traffic and Related Self-Driven Many-Particle Systems," *Reviews of Modern Physics* 73, 1067 (2001), and references therein.

46   **an intrinsic instability of the traffic itself:** These intrinsic jams can cost us real money—a 2011 "Urban Mobility Report" from the Texas Transportation Institute estimated the economic impact of traffic jams on the U.S. economy to be roughly $100 billion per year, equivalent to $750 per commuter.

46   **described as a collective phenomenon:** B. S. Kerner and P. Konhauser, "Cluster Effect in Initially Homogeneous Traffic Flow," *Physical Review E* 48, 2335 (1993); Kai Nagel and Maya Paczuski, "Emergent Traffic Jams," *Physical Review E* 51, 2909 (1995); H. Y. Lee, H.-W. Lee, and D. Kim, "Origin of Synchronized Traffic Flow on Highways and Its Dynamic Phase Transition," *Physical Review Letters* 81, 1130 (1998); B. S. Kerner, "Experimental Features of Self-Organization in Traffic Flow," *Physical Review Letters* 81, 3797 (1998).

47   **Two factors play a key role:** Robert E. Chandler, Robert Herman, and Elliott W. Montroll, "Traffic Dynamics: Studies in Car Following," *Operations Research* 6, 163 (1958).

47   **An analogous situation:** While there is considerable overlap between the dynamics of granular media and

traffic flow (see, for example, the proceedings of the International Workshop on Traffic and Granular Flow, held every two years for the past two decades), the equations describing a sandpile's instability differ from those modeling the spontaneous formation of traffic jams.

47   **one person at the front edge:** Takashi Nagatani, "Traffic Jams Induced by Fluctuation of a Leading Car," *Physical Review E* 61, 3534 (2000).

48   **drive at a steady, uniform speed:** Junfang Tian, Rui Jiang, Geng Li, Martin Treiber, Bin Jia, and Chenqiang Zhu, "Improved 2D Intelligent Driver Model in the Framework of Three-Phase Traffic Theory Simulating Synchronized and Concave Growth Patterns of Traffic Oscillations," *Transportation Research Part F: Traffic Psychology and Behavior* 41, 55 (2016).

50   **"radar" is an acronym:** See, for example, J. C. Toomay and Paul J. Hannen, *Radar Principles for the Non-Specialist,* 3rd ed. (SciTech Publishing, 2004).

50   **traced back to the proximity fuse:** Barry Parker, *The Physics of War: From Arrows to Atoms* (Prometheus Books, 2014), 241–42.

52   **Your remote entry system:** Mark Fischetti, "Open Sesame," *Scientific American,* January 2005.

53   **Your radio tuner:** Bloomfield, *How Everything Works,* 430.

54   **Couldn't someone with a receiver:** Matt Lake, "How It Works; Remote Keyless Entry: Staying a Step Ahead of Car Thieves," *New York Times,* June 7, 2001.

CHAPTER THREE: YOU GO TO THE DOCTOR

57   **An elevator is essentially a pulley:** David Macaulay, *The Way Things Work* (Houghton Mifflin Company, 1988), 65.

59   **Speed is mostly dependent on the motor:** Lauren Anderson, "How It Works: World's Fastest Elevator," *Popular Science,* April 2012.

59   **The main rule for elevator safety:** Ibid.

61   **"governor" in the rotating pulley:** Louis Bloomfield, *How Everything Works,* 5th ed. (John Wiley & Sons, 2008), 123.

64   **The website address you type:** Andrew Blum, *Tubes: A Journey to the Center of the Internet* (HarperCollins, 2012), 29–30, 49–55.

64   **The difference between analog and digital:** Barry M. Lunt, *Marvels of Modern Electronics: A Survey* (Dover, 2013), 173–75.

65   **A more reliable way to transmit information:** See, for example, John R. Pierce, *An Introduction to Information Theory: Symbols, Signals and Noise,* 2nd rev. ed. (Dover, 1980).

66   **a sequence of closely spaced dots:** See, for example, Jiun-Haw Lee, David N. Liu, and Shin-Tson Wu, *Introduction to Flat Panel Displays* (John Wiley & Sons, 2009).

67   **The old-school glass thermometers:** Mark W. Zemansky, *Heat and Thermodynamics,* 5th ed. (McGraw-Hill, 1968), 9–14.

69   **Any system can be used as a thermometer:** Guy K. White, *Experimental Techniques in Low-Temperature*

Physics, 3rd ed. (Oxford Science Publications, 1979), 83–122.

70   **When two dissimilar metals:** D. K. C. MacDonald, *Thermoelecticity: An Introduction to the Principles* (Dover Books, 2006), 12–15; White, *Experimental Techniques in Low-Temperature Physics*, 117–22. See James Kakalios, *The Amazing Story of Quantum Mechanics* (Gotham, 2010), 256, for a discussion of thermocouples.

72   **In an x-ray tube:** Robert Eisberg and Robert Resnick, *Quantum Physics of Atoms, Molecules, Solids, Nuclei and Particles*, 2nd ed. (John Wiley & Sons, 1985), 40–43.

73   **As x-rays move into a solid:** See, for example, William H. Zachariasen, *Theory of X-Ray Diffraction in Crystals* (Dover, 1967).

76   **"computer-aided tomography":** Jerrold T. Bushberg, J. Anthony Seibert, Edwin M. Leidholdt Jr., and John M. Boone, *The Essential Physics of Medical Imaging*, 3rd ed. (LLW, 2011), 312–75.

78   **In ultrasound imaging:** See, for example, Marveen Craig, *Essentials of Sonography and Patient Care,* 3rd ed. (Saunders, 2012); and James A. Zagzebski, *Essentials of Ultrasound Physics* (Mosby, 1996).

82   **each have a small, built-in magnetic field:** Eisberg and Resnick, *Quantum Physics of Atoms, Molecules, Solids, Nuclei and Particles*, 434–37.

83   **first recognized by Albert Einstein back in 1905:** Barry Parker, *Quantum Legacy: The Discovery That Changed Our Universe* (Prometheus Books, 2002), 46–50.

84   **The dominant MRI signal:** See, for example, Dominik Weishaupt, Victor D. Koechli, and Borut Marincek,

How Does MRI Work? An Introduction to the Physics and Function of Magnetic Resonance Imaging (Springer, 2008). A description of the physics behind magnetic resonance imaging is in Kakalios, *The Amazing Story of Quantum Mechanics,* 227–33.

86   **One recent study:** Simon Walker-Samuel et al., "*In Vivo* Imaging of Glucose Uptake and Metabolism in Tumors," *Nature Medicine* 19, 1067 (2013).

CHAPTER FOUR: YOU GO TO THE AIRPORT

88   **antique technology of carbon paper:** See "Carbon Paper" on the *How Products Are Made* website, http://www.madehow.com/Volume-1/Carbon-Paper.html #ixzz4IkE9h9fE; and Kevin Laurence, "The *Exciting* History of Carbon Paper!," http://www.kevinlaurence.net/essays/cc.php.

89   **magnetic stripe on your credit card:** Robert C. O'Handley, *Modern Magnetic Materials: Principles and Applications* (John Wiley & Sons, 2000), 674–77; Barry M. Lunt, *Marvels of Modern Electronics: A Survey* (Dover, 2013), 150–52; Thomas Norman, *Electronic Access Control* (Butterworth-Heinemann, 2011), 53–54.

89   **Since binary information is the foundation:** T. R. Reid, *The Chip: How Two Americans Invented the Microchip and Launched a Revolution* (Simon & Schuster, 1985), 121–27. For a primer on how to use binary to count to 31 using the fingers of one hand, see *The Unbeatable Squirrel Girl,* no. 11, written by Ryan North and drawn by Jacob Chabot and Erica Henderson (Marvel Comics, October 2016).

91n   **using a "read head":** Jagadeesh S. Moodera, Guo-Xing Miao, and Tiffany S. Santos, "Frontiers in Spin-Polarized Tunneling," *Physics Today,* April 2010.

92   **often termed an RFID:** Roy Want, "RFID—A Key to Automating Everything," *Scientific American*, August 2008.

94   **"resistive" sensing:** Geoff Walker, "A Review of Technologies for Sensing Location on the Surface of a Display," *Journal of the Society for Information Display* 20, 413 (2012).

95   **device called a "capacitor":** Paul A. Tipler and Gene Mosca, *Physics for Scientists and Engineers*, 6th ed. (W. H. Freeman, 2007), 802–806.

95   **When the charge stored on one:** Walker, "A Review of Technologies."

96   **a special alloy, indium tin oxide:** H. J. J. van Boort and R. Groth, "Low Pressure Sodium Lamps with Indium Oxide Filter," *Phillips Technical Review* 29, 17 (1968); Mamoru Mizuhashi, "Electrical Properties of Vacuum-Deposited Indium Oxide and Indium Tin Oxide Films," *Thin Solid Films* 70, 91 (1980).

98   **Some touch screens use light:** Stuart F. Brown, "Hands-On Computing," *Scientific American*, July 2008.

99   **"total internal reflection":** Lunt, *Marvels of Modern Electronics: A Survey*, 202–207.

99   **the basis of fiber-optic cable:** In a fiber-optic cable, the light travels down a cylindrical core that is surrounded by an outer shell (called the "cladding") of another material with a lower index of refraction. If these indexes are chosen carefully, then the light will be reflected at the core/cladding interface, regardless of the angle it makes with the surface. Light that travels in a straight-line beam, as from a laser, will bounce

off the cladding and continue its propagation down the core fiber, no matter how many twists or turns the cable makes. See Ben G. Streetman and Sanjay Banerjee, *Solid State Electronic Devices,* 5th ed. (Prentice Hall, 2000), 392–94.

100   **handheld metal detectors:** Marshall Brain, *MORE How Stuff Works* (John Wiley & Sons, 2002), 291–93.

101   **walk-through detectors:** David Macaulay, *The Way Things Work* (Houghton Mifflin Company, 1988), 323.

103   **scanners at airports use "millimeter waves":** R. Appleby, "Passive Millimetre-Wave Imaging and How It Differs from Terahertz Imaging," *Philosophical Transactions of the Royal Society of London A* 362, 379 (2004).

104   **electromagnetic wave with a wavelength:** Paul A. Tipler and Gene Mosca, *Physics for Scientists and Engineers* vol. 1, 6th ed. (W. H. Freeman, 2007), 1041.

106   **insulators or semiconductors as an old-style movie theater:** James Kakalios, *The Amazing Story of Quantum Mechanics* (Gotham, 2010), 175–78.

107   **In some detectors:** John Rowlands and Safa Kasap, "Amorphous Semiconductors Usher in Digital X-Ray Imaging," *Physics Today,* November 1997; Yoshihiro Izumi and Yasukuni Yamane, "Solid-State X-Ray Imagers," *MRS Bulletin,* November 2002; Martin Niki, "Scintillation Detectors for X-rays," *Measurement Science and Technology* 17, R37 (2006).

107   **absorption of either x-ray or visible-light:** Rowlands and Kasap, "Amorphous Semiconductors Usher in Digital X-Ray Imaging."

110 **A molecule of trinitrotoluene:** See, for example, G. A. Eiceman, Z. Karpas, and Herbert H. Hill Jr., *Ion Mobility Spectrometry*, 3rd ed. (CRC Press, 2016).

111 **"ion mobility spectroscopy":** Abu B. Kanu, Prabha Dwivedi, Maggie Tam, Laura Matz, and Herbert H. Hill Jr., "Ion Mobility-Mass Spectrometry," *Journal of Mass Spectrometry* 43, 1 (2008).

CHAPTER FIVE: YOU TAKE A FLIGHT

116 **Lithium, like its alkaline cousins:** Louis Bloomfield, *How Everything Works*, 5th ed. (John Wiley & Sons, 2008), 632–45; and see Seth Fletcher, *Bottled Lightning: Superbatteries, Electric Cars and the New Lithium Economy* (Hill and Wang, 2011).

116 **As conventional alkaline batteries age:** Isidor Buchmann, *Batteries in a Portable World: A Handbook on Rechargeable Batteries for Non-Engineers*, 4th ed. (Cadex Electronics, 2016); Thomas Reddy, *Linden's Handbook of Batteries*, 4th ed. (McGraw-Hill Education, 2010).

117 **Years ago, batteries in automobiles:** Henry Schlesinger, *The Battery: How Portable Power Sparked a Technological Revolution* (Smithsonian Books, 2010), 173–75; Fletcher, *Bottled Lightning*, 13–17.

120 **Hot air balloons and blimps:** David Macaulay, *The Way Things Work* (Houghton Mifflin Company, 1988), 103, 112–13.

121 **An airplane's wings:** Bloomfield, *How Everything Works*, 168–71.

123 **A jet engine may not look like it:** Ibid., 175–76.

125 **silicon chip sandwich:** Mark Johnson, *Photodetection and Measurement: Maximizing Performance in Optical*

Systems (McGraw-Hill Education, 2003), 1–11; Ben G. Streetman and Sanjay Banerjee, *Solid State Electronic Devices*, 5th ed. (Prentice Hall, 2000), 379–86.

126   **"bucket brigade principle":** See, for example, James R. Janesick, *Scientific Charge-Coupled Devices* (SPIE Press, 2001).

131   **multitude of computer memory banks:** Andrew Blum, *Tubes: A Journey to the Center of the Internet* (Harper-Collins, 2012), 259; Prashant Gupta, A. Seetharaman, and John Rudolph Raj, "The Usage and Adoption of Cloud Computing by Small and Medium Businesses," *International Journal of Information Management* 33, 861 (2013).

132   **If the heat is not removed:** Vaclav Smil, *Energies: An Illustrated Guide to the Biosphere and Civilization* (MIT Press, 1999), 200.

132   **Heat management is a major issue:** Matt McKinney, "Selling the Cold, Minnesota's Tech Community Welcomes Data Centers," *Minneapolis Star Tribune*, December 13, 2014.

132   **data centers run by Google:** James Glanz, "Google Details, and Defends, Its Use of Electricity," *New York Times*, September 8, 2011; though Google's carbon footprint is scheduled to change: Quentin Hardy, "Google Says Its Data Centers Will Run Entirely on Renewable Energy by 2017," *New York Times*, December 7, 2016.

134   **Sound is manifested by density variations:** Paul A. Tipler and Gene Mosca, *Physics for Scientists and Engineers*, vol. 1, 6th ed. (W. H. Freeman, 2007), 502–503.

136   **Noise-canceling headphones record:** S. J. Elliott and P. A. Nelson, "Active Noise Control," *IEEE Signal Pro-*

cessing Magazine, October 1993; Scott D. Snyder, *Active Noise Control Primer (Modern Acoustics and Signal Processing)* (Springer, 2000), 3–5.

137 **The constituents of atoms:** Robert Eisberg and Robert Resnick, *Quantum Physics of Atoms, Molecules, Solids, Nuclei and Particles,* 2nd ed. (John Wiley & Sons, 1985), 272–78, 434–37.

138 **magnetically levitating trains:** David Scott, "At Last, Maglev Goes Public: Britain's Flying Train," *Popular Science,* October 1984; Scott R. Gourley, "Track to the Future," *Popular Mechanics,* May 1998; see, for example, Hyung-Suk Han and Dong-Sung Kim, *Magnetic Levitation: Maglev Technology and Applications* (Springer Tracts on Transportation and Traffic, 2016).

CHAPTER SIX: YOU GIVE A BUSINESS PRESENTATION
142 **To make a transistor:** Barry M. Lunt, *Marvels of Modern Electronics: A Survey* (Dover, 2013), 23–32.

144 **a third metal plate, buried within the insulating glass:** D. Kahng and S. M. Sze, "A Floating-Gate and Its Application to Memory Devices," *The Bell System Technical Journal* 46, 1288 (1967); R. Bez, E. Camerlenghi, A. Modelli, and A. Visconti, "Introduction to Flash Memory," *Proceedings of the IEEE* 91, 489 (2003). The physics of transistors with and without a floating gate is considered in James Kakalios, *The Amazing Story of Quantum Mechanics* (Gotham, 2010), 216–18.

145 **discs with magnetic regions:** Robert C. O'Handley, *Modern Magnetic Materials: Principles and Applications* (John Wiley & Sons, 2000), 674–721; Lunt, *Marvels of Modern Electronics: A Survey,* 153–56.

145n   **a quantum mechanical process termed "tunneling":** Robert Eisberg and Robert Resnick, *Quantum Physics of Atoms, Molecules, Solids, Nuclei and Particles,* 2nd ed. (John Wiley & Sons, 1985), 205–209.

146   **A photocopier starts with a semiconductor:** H. Richard Crane, *How Things Work* (American Association of Physics Teachers, 1996), 18–19.

147   **called "toner":** Dan A. Hays, "How Does a Photocopier Work?" *Scientific American,* March 2003.

148   **The seats in the lower-energy orchestra:** Kakalios, *The Amazing Story of Quantum Mechanics,* 198–201.

148n   **To quote David Owen:** David Owen, "Copies in Seconds," *The Atlantic Monthly,* February 1986.

149   **One form of ink-jet printing:** Marshall Brain, *MORE How Stuff Works* (John Wiley & Sons, 2002), 232–33.

151   **made up of an amorphous semiconductor:** See, for example, S. R. Elliott, *Physics of Amorphous Materials* (Longman Scientific & Technical, 1983).

151   **A crystalline seed is dipped:** Ben G. Streetman and Sanjay Banerjee, *Solid State Electronic Devices,* 5th ed. (Prentice Hall, 2000), 12–16.

153   **A liquid crystal is a true fluid:** See, for example, Peter J. Collings, *Liquid Crystals: Nature's Delicate Phase of Matter* (Princeton University Press, 1990).

154   **flat panel displays:** See, for example, Jiun-Haw Lee, David N. Liu, and Shin-Tson Wu, *Introduction to Flat Panel Displays* (John Wiley & Sons, 2009); and Robert H. Chen, *Liquid Crystal Displays: Fundamental Physics and Technology* (John Wiley & Sons, 2011).

157 **A diode consists of two semiconductors:** John P. McKelvey, *Solid State and Semiconductor Physics* (Harper & Row, 1966), 390–98, 408–16.

160 **Shine light on the semiconductor:** Charles Kittel, *Introduction to Solid State Physics*, 7th ed. (John Wiley & Sons, 1996), 570–74.

161 **This is a light-emitting diode (LED):** Nick Holonyak, "From Transistors to Lasers to Light-Emitting Diodes," *MRS Bulletin* 30, 509 (2005); Chris Woodford, *Atoms Under the Floorboard* (Bloomsbury Sigma, 2015), 159–61. The physics of light-emitting diodes and laser diodes is described in Kakalios, *The Amazing Story of Quantum Mechanics*, 184–87, 207–209.

161 **To turn an LED into a semiconductor laser:** Winston Kock, *Lasers and Holography: An Introduction to Coherent Optics*, 2nd ed. (Dover, 1981).

162 **The first semiconductor laser diodes:** Louis Bloomfield, *How Everything Works*, 5th ed. (John Wiley & Sons, 2008), 26–32, 475.

164 **Microphones in early models of landline telephones:** See, for example, Glen Ballou, ed., *Electroacoustic Devices: Microphones and Loudspeakers* (Focal Press, 2009); and M. D. Fagen, ed., *A History of Engineering and Science in the Bell System: The Early Years (1875–1925)* (Bell Telephone Laboratories, 1975).

165 **In a condenser microphone:** Alexander Case, "The Vocal Microphone: Technology and Practice," *Physics Today*, March 2016.

166 **"ribbon microphones":** Ibid.

167 **The "electret microphone":** Leo L. Beranek and Tim Mellow, *Acoustics: Sound Fields and Transducers* (Academic Press, 2012), 397–400.

169   **roughly a few electron-Volts:** See, for example, Roy McWeeny, *Coulson's Valence* (Oxford University Press, 1979).

171   **Microwave ovens use electromagnetic waves:** Michael Vollmer, "Physics of the Microwave Oven," *Physics Education* 39, 74 (2004).

172   **This force is called the Strong Force:** A. Das and T. Ferbel, *Introduction to Nuclear and Particle Physics,* 2nd ed. (World Scientific, 2003), 45–50.

173   **The predominant forms of radiation:** George Gamow, *The Atom and Its Nucleus* (Prentice Hall, 1961), 78–81.

174   **One way to eliminate the bacteria:** Jozsef Farkas and Csilla Mohacsi-Farkas, "History and Future of Food Irradiation," *Trends in Food Science & Technology* 22, 121 (2011).

175   *irradiated food is not itself radioactive:* There is a phenomenon termed "induced radioactivity," where a stable nucleus can be transmuted into an unstable and hence radioactive nucleus. However, this occurs only if the nucleus is exposed to an external source of neutrons, and does not occur with the gammas, x-rays, and beta particles used in food sterilization.

CHAPTER SEVEN: YOU GO TO A HOTEL

179   **accelerometers that exploit:** Jon S. Wilson, *Sensor Technology Handbook,* vol. 1 (Elsevier, 2005), 137–53; S. Beeby, G. Ensell, M. Kraft, and N. White, *MEMS Mechanical Sensors* (Artech House Inc., 2004), 175–95; O. Sidek, M. N. Mat Nawi, and M. A, Miskam, "Analysis of Low-g Capacitive Cantilever-Mass Micro-Machined

Accelerometers," *International Joural of Engineering & Technology* 10, 133 (2010).

181 **To record your pulse:** See, for example, John G. Webster, ed., *Design of Pulse Oximeters* (CRC Press, 1997).

182 **The motion sensor detects this infrared light:** Jacob Fraden, *Handbook of Modern Sensors: Physics, Designs and Applications*, 4th ed. (Springer, 2010), 95–100, 487–91.

184 **Roughly half of the energy:** Vaclav Smil, *Energy in Nature and Society: General Energetics of Complex Systems* (MIT Press, 2008), 124–27.

185 **"pyroelectric detector":** See, for example, A. K. Batra and M. D. Aggarwal, *Pyroelectric Materials: Infrared Detectors, Particle Accelerators and Energy Harvesters* (SPIE Press, 2013).

186 **"active" motion sensors:** Fraden, *Handbook of Modern Sensors*, 249–54.

187 **passive card-entry system:** See, for example, Robert N. Reid, *Facility Manager's Guide to Security: Protecting Your Assets* (Fairmont Press, 2005); Thomas Norman, *Electronic Access Control* (Butterworth-Heinemann, 2011).

189 **Active proximity cards:** Ibid.

190 **To make a remote control device:** See, for example, Julia Layton, "How Remote Controls Work," November 10, 2005, http://electronics.howstuffworks.com/remote -control.htm.

191 **band structure of solids:** Charles Kittel, *Introduction to Solid State Physics*, 7th ed. (John Wiley & Sons, 1996), 173–94.

192   **what we know about molecules:** Robert Eisberg and Robert Resnick, *Quantum Physics of Atoms, Molecules, Solids, Nuclei and Particles,* 2nd ed. (John Wiley & Sons, 1985), 422–32.

193n   **"greenhouse gas":** Vaclav Smil, *Energy in Nature and Society: General Energetics of Complex Systems* (MIT Press, 2008), 33–34.

194   **"organic light-emitting diodes" (OLEDs):** Eliav I. Haskal, Michael Buchel, Paul C. Duineveld, Aad Sempel, and Peter van de Weijer, "Passive-Matrix Polymer Light-Emitting Displays," *MRS Bulletin,* November 2002; Jiun-Haw Lee, David N. Liu, and Shin-Tson Wu, *Introduction to Flat Panel Displays* (John Wiley & Sons, 2009).

197   **blue light OLED:** Hungshin Fu, Yi-Ming Cheng, Pi-Tai Chou, and Yun Chi, "Feeling Blue? Blue Phosphors for OLEDs," *Materials Today* 14, 472 (2011); Jeong-A Seo, Sang Kyu Jeon, Myoung Seon Gong, Jun Yeob Lee, Chang Ho Noh, and Sung Han Kim, "Long Lifetime Blue Phosphorescent Organic Light-Emitting Diodes with an Exciton Blocking Layer," *Journal of Materials Chemistry C* 3, 4640 (2015).

200   **the important issue is** *energy density*: Vaclav Smil, *Energy: A Beginner's Guide* (Oneworld Publications, 2006), 15–16. The impracticalities of flying cars using conventional fuel is discussed in James Kakalios, *The Physics of Superheroes: Spectacular Second Edition* (Gotham Books, 2009), 151–52.

200   **how much stored chemical potential energy can you carry per pound of fuel?:** A pound of wood contains approximately 6.8 million Joules of energy. (Recall

that one way that energy can be measured is in Joules, where an average person taking a jaunty stroll will have roughly fifty Joules of kinetic energy.) While this may sound like a lot, coal and natural gas have over twice the energy per pound, with gasoline having 20.5 million Joules per pound. There are fuels with even higher energy densities, such as hydrogen gas (consisting of two hydrogen atoms bound together). At 52 million Joules per pound, hydrogen has an even higher energy density than gasoline, due to its unique combination of large potential energy stored in its chemical bonds, and the lowest weight of any molecule. However, gasoline is easier (and hence cheaper) to store and transport than hydrogen gas, and once the decision was made roughly a hundred years ago to use a particular fuel in the early-model internal combustion engines, it is not a simple matter to change, even with hydrogen's promise of over twice the energy bang for the buck. (The fear of an actual bang for a buck is another issue with all-hydrogen cars, though a gasoline fire can be just as, if not more, dangerous.) Smil, *Energy: A Beginner's Guide*, 16.

200   **If two deuterium nuclei are fused:** Charles Seife, *Sun in a Bottle* (Viking, 2008), 139–41.

202   **the best energy densities:** See, for example, B. E. Conway, *Electrochemical Supercapacitors: Scientific Fundamentals and Technological Applications* (Springer, 1999); and Hector D. Abruna, Yasuyuki Kiya, and Jay C. Henderson, "Batteries and Electrochemical Capacitors," *Physics Today* 61, 43 (2008).

204   **Such carbon is termed "activated":** Elzbieta Frackowiak and Francois Beguin, "Carbon Materials for the Elec-

trochemical Storage of Energy in Capacitors," *Carbon* 39, 937 (2001).

204   **are termed "ultra-capacitors," or "super-capacitors":** Joel Schindall, "The Charge of the Ultra-Capacitors," *IEEE Spectrum* 44, 42 (2007); Phillip Ball, "A Capacity for Change," *MRS Bulletin* 37, 1000 (2012).

Figure 1

Sketch of a pendulum with a positively charged bob swinging back and forth. *(a)* When bob is at its highest point on the left, there is no kinetic energy, but the potential energy is maximum, and there is no current (the charge is not moving). *(b)* Bob at its lowest point has a large kinetic energy, but no potential energy. There is a large electrical current to the right. *(c)* Bob at its highest point, at the right, again has no kinetic energy, but the potential energy is again maximum, and there is no current. *(d)* When bob is moving to the left at the midpoint of its swing, there is some kinetic and some potential energy, and there is an electrical current to the left, which is changing as the charge is speeding up and also changing direction.

Figure 2

Sketch of a radio transmitter and receiver. An electron current oscillates up and down the transmitter *(top)* and generates an electromagnetic wave with the same frequency as the current's oscillation. When this wave intersects the metal antenna *(bottom),* it induces electrons in the metal to oscillate with the same frequency as the wave. This physical principle is employed in an E-ZPass system, a garage door opener, and a keyless entry remote.

Figure 3

Sketch of an x-ray scan. The top drawing shows an x-ray tube, where an electron current is drawn from the coil on the left toward the plate on the right. The sudden deceleration of the electron current leads to the creation of high-frequency electromagnetic waves termed "x-rays." When these x-rays are directed to your ankle, they interact with the atoms in your body. The more electrons the atom has, the more strongly it scatters x-rays from their original path. The x-rays are deflected by the bone, but pass through the tissue. The detector plate thus records a high x-ray flux except where bones are present.

Figure 4

Sketch of a parallel-plate capacitor as used in a touch screen display. The two plates of a transparent conductor are separated by a thin insulator. Excess electrical charges on your fingertip change the charge density on the capacitor plates, at the location touched.

Figure 5

Sketch of the cross-section of an airplane wing, showing the flow of air past the wing. The wing is angled so that air striking the bottom surface is deflected downward, and due to

Newton's third law (forces come in pairs), the air exerts an equal and opposite upward force (termed "lift") on the bottom of the wing.

Figure 6.1

Sketch of a liquid crystal display (LCD) pixel. In the top drawing, unpolarized white light passes through a vertical polarizer that passes light with its oscillating electric field in the vertical plane. The polarization of the light twists along with the liquid crystal molecules, so the light is able to pass through the second polarizer, which is oriented in the horizontal position. In the bottom drawing, an electric field is applied across the cell using a capacitor with transparent conductors for plates. The liquid crystal now shifts into a phase where it does not rotate into the horizontal plane, so the polarized light is unable to pass through the second polarizer, and the pixel is now dark.

Figure 6.2

Sketch of a p–n junction semiconductor—a diode. In the upper drawing there is a built-in electric field across the boundary between the semiconductor with an excess of electrons (n-type) on the left and the semiconductor with an excess of holes (p-type) on the right. If a current is forced through the junction, in opposition to this built-in electric field, then at the junction, the electrons lower their energy by falling into the holes, releasing photons of light—this is a light-emitting diode. In the lower drawing, the reverse process is shown. Light is incident on the device, and electrons are promoted from the lower filled band to the higher energy empty band, leaving holes behind in the lower band of states. The built-in electric field of the diode now pushes the electrons to the left and the holes to the right—a current results when light is absorbed. This is a photodiode, or a solar cell.

Figure 7

A flying car, with four small jet engines at the wheel locations. The challenge of building such a vehicle is carrying enough fuel with a high enough energy density that the trip lasts more than a few minutes.

INDEX

Page numbers in *italics* refer to illustrations. Page numbers beginning with 211 refer to endnotes.

ABOUT THE AUTHOR

JAMES KAKALIOS is the Taylor Distinguished Professor in the School of Physics and Astronomy at the University of Minnesota and the author of the bestselling *The Physics of Superheroes*. He is a condensed matter experimentalist whose research concerns amorphous/nanocrystalline composite semiconductor materials and fluctuation phenomena in neurological systems. He is the recipient of a regional Emmy Award, and his efforts at public outreach and science communication have been recognized by awards from the American Institute of Physics and the American Association for the Advancement of Science.